Experimental Simulation and Characterization of Radiation Damage in Materials

Experimental Simulation and Characterization of Radiation Damage in Materials

Editor

Vladimir Krsjak

MDPI • Basel • Beijing • Wuhan • Barcelona • Belgrade • Manchester • Tokyo • Cluj • Tianjin

Editor
Vladimir Krsjak
Slovak University of Technology in Bratislava
Slovakia

Editorial Office
MDPI
St. Alban-Anlage 66
4052 Basel, Switzerland

This is a reprint of articles from the Special Issue published online in the open access journal *Materials* (ISSN 1996-1944) (available at: https://www.mdpi.com/journal/materials/special_issues/simulation_radiation).

For citation purposes, cite each article independently as indicated on the article page online and as indicated below:

LastName, A.A.; LastName, B.B.; LastName, C.C. Article Title. *Journal Name* **Year**, *Volume Number*, Page Range.

ISBN 978-3-0365-6752-5 (Hbk)
ISBN 978-3-0365-6753-2 (PDF)

Cover image courtesy of Pavol Noga

Contents

About the Editor

Vladimir Krsjak

Vladimir Krsjak, PhD, is a research associate at the Institute of Nuclear and Physical Engineering (INPE) and the Advanced Technology Research Institute (ATRI) of the Slovak University of Technology in Bratislava, Slovakia (STU). He received his PhD in nuclear power engineering from STU in 2008 on the non-destructive characterization of radiation effects in nuclear materials. He continued as a post-doc researcher at the Joint Research Centre of the European Commission (Netherlands) in the Safety of Future Nuclear Reactors Unit. In 2012 he joined the Laboratory for Nuclear Materials at the Swiss Paul Scherrer Institute. During his postdoctoral appointments, he has worked with leading experts in radiation materials science. He took part in several international irradiation experiments involving thermal and fast reactor irradiation, spallation neutron target irradiation and ion bombardment, focusing on the potential life extension of the current nuclear reactor fleet and material challenges for the future of nuclear power. His current research interests include accelerated radiation ageing of nuclear structural materials and semiconductors with an experimental focus on advanced microstructural characterization methods.

Preface to "Experimental Simulation and Characterization of Radiation Damage in Materials"

Developing new materials for the next generation of advanced nuclear technologies brings about the need for suitable irradiation experiments and reliable/reproducible post-irradiation examination (PIE). The standardized testing of activated materials in hot-cell facilities might still be considered unavoidable for nuclear power reactor development. However, exposure to fission neutron radiation cannot adequately simulate the radiation environments of nuclear fusion or spallation neutron sources. For this reason, considerable efforts are being devoted to using ion bombardment as a surrogate for neutron irradiation and developing new, innovative methods for material characterization. The use of particle accelerators to simulate harsh radiation environments provides an easily and well controlled environment that enables us to isolate and study the effects of temperature and displacement damage rate or to effectively simulate transmutation reactions in materials such as helium or hydrogen production.

This book aims to advance the current knowledge in ion irradiation studies and the accelerated ageing of materials. The collected papers address various areas in the field, from the fundamentals of ion beam modification to innovative experimental characterization of the materials and related theoretical modelling. The present research contributes primarily to the understanding of the applied aspects of ion bombardment as a surrogate for neutron irradiation and a tool for experimental simulation of harsh radiation environments. At the same time, this Special Issue features several exciting studies on the microstructural characterization and radiation tolerance of materials across a broad spectrum of scientific and industrial areas.

Vladimir Krsjak
Editor

materials

Article

High-Fluence Multi-Energy Ion Irradiation for Testing of Materials

Pavol Noga [1,*], Zoltán Száraz [1], Matej Kubiš [1], Jozef Dobrovodský [1], Filip Ferenčík [1], Róbert Riedlmajer [1] and Vladimir Krsjak [1,2]

1 Slovak University of Technology in Bratislava, Faculty of Materials Science and Technology in Trnava, Advanced Technologies Research Institute, Jána Bottu 25, 91724 Trnava, Slovakia
2 Slovak University of Technology in Bratislava, Faculty of Electrical Engineering and Information Technology, Institute of Nuclear and Physical Engineering, Ilkovičova 3, 81219 Bratislava, Slovakia
* Correspondence: pavol.noga@stuba.sk

Abstract: Structural materials of the new generation of nuclear reactors, fission as well as fusion, must often cope with high production rates of transmutation helium. Their testing hence requires either a powerful source of fast neutrons or a high-fluence ion-irradiation facility providing sufficient amounts of high-energy helium to investigate its effect on the material. Most ion irradiation studies, however, concentrate on basic effects such as defect evolution or bubble swelling in narrow near-surface regions modified by ion bombardment. Studies on bulk samples with a relatively thick implanted region, which would enable, for instance, micromechanical testing, are underrepresented. This gap might be filled by high-fluence multi-energy ion irradiations modifying several tens of micrometres of the investigated substrate. High-energy ion accelerators providing reasonable currents with energies of tens of MeV are rarely employed in such studies due to their scarcity or considerable beamtime costs. To contribute to this field, this article reports a unique single-beam He implantation experiment aimed at obtaining quasi-uniform displacement damage across >60 μm with the He/dpa ratio roughly one order of magnitude above the typical spallation neutron target irradiation conditions. Some technical aspects of this irradiation experiment, along with recent developments and upgrades at the 6 MV Tandetron accelerator of the Slovak university of technology in Bratislava, are presented.

Keywords: high-energy ion irradiation; nuclear materials; transmutation helium

Citation: Noga, P.; Száraz, Z.; Kubiš, M.; Dobrovodský, J.; Ferenčík, F.; Riedlmajer, R.; Krsjak, V. High-Fluence Multi-Energy Ion Irradiation for Testing of Materials. *Materials* **2022**, *15*, 6443. https://doi.org/10.3390/ma15186443

Academic Editor: Gianpiero Buscarino

Received: 24 August 2022
Accepted: 13 September 2022
Published: 16 September 2022

Publisher's Note: MDPI stays neutral with regard to jurisdictional claims in published maps and institutional affiliations.

1. Introduction

The growing demand for more energy with simultaneous efforts towards carbon-free energy production increases the importance of the nuclear energy sector. In the recent years there has been an interest to develop fourth-generation (GEN IV) fission reactors, small modular reactors, and fusion reactors. Compared with the current reactor conditions, materials in advanced nuclear systems need to withstand higher temperatures, more corrosive coolants, and prolonged high-energy neutron irradiation. While the operating temperature of commercial light water reactors does not exceed 350 °C, the six concepts of the future fission systems, proposed within the Generation IV international forum, will operate in a temperature range of 350–1000 °C. The foreseen end of lifetime damage levels are up to ~200 dpa [1,2]. In prototype fusion devices the damage of 150–200 dpa in the replaceable structures will be caused by the 14 MeV neutrons generated during D-T fusion reaction. The demonstration fusion power plant DEMO is expected to operate from 300 to 1000 °C [3].

Higher neutron flux and harder neutron spectra, together with increased temperatures, call for the development of new radiation-tolerant structural materials. For the deployment of the abovementioned new systems, it is crucial to understand how radiation degrades these materials and how various parameters affect their irradiation response. The materials development is a challenging, lengthy process as it needs to go through several steps and iterations, and the neutron irradiation to significant exposures takes a long time. Other limiting factors are the decreased availability of suitable materials test reactors and the very high cost of neutron irradiation. To shorten the 40–50-year process, materials development with a focus on high-fluence irradiation conditions uses the best alternative technique available to date: charged particle irradiation, at least until facilities such as the High Flux Accelerator-Driven Neutron Facility (HF-ADNeF) [4,5] or the much more powerful IFMIF-DONES facility [6] capable of producing a neutron spectrum very close to the conditions in fusion reactors and at high fluxes become operational. Yet, operation of the latter is planned no earlier than 2033 [7].

Ion irradiation is widely employed to investigate radiation-induced microstructural changes and the resultant material damage. Ion implantation enables much faster damage accumulation in comparison with nuclear reactors, and the dose corresponding to several years of neutron irradiation can be reached in a matter of hours or days and is therefore considered as the best available surrogate to neutron irradiation known to date. However, the penetration of the accelerated ions into the material is limited. The irradiation depth achieved during low energy proton or heavy ion irradiation (up to 100 MeV) is in the order of submicron to a few micrometres for common metals used in nuclear materials [8]. Therefore, the damage is confined into a thin near-surface region, in the order of μm, and the resulting displacement profile is graded. This makes it difficult to determine and evaluate the mechanical properties of ion-irradiated materials and limits the testing and investigation to the nanoscale level. Most ion irradiation facilities for materials research provide relatively low-energy protons or heavy ions (2–5 MeV), restricting the investigation of irradiation induced changes in mechanical properties under irradiation. Moreover, most such investigations limit themselves to a single implantation/irradiation step.

For materials development and qualification, engineering data such as strength, ductility, toughness, etc. are needed. To extract bulk properties, the range limitation in ion implantation experiments needs to be overcome. Increasing the irradiated layer thickness to 50–100 μm or beyond will make micro-mechanical testing [9] feasible.

Obtaining such "bulk" volumes of ion-beam modified material requires employing high-energy ion accelerators with powerful ion sources. Ion bombardment using light particles with energy > 10 MeV, however, leads to considerable amounts of gamma and neutron radiation during the beam-on time. This must be considered in the accelerator facility layout and resolved by sufficient radiation shielding.

One of the few setups dedicated especially to "thick layer" (in sense of the above) irradiation is located at MIT [8], based on a cyclotron providing 10–30 MeV proton irradiations and a full mechanical tensile test stage with 100–300 μm-thick samples (proton ranges at these energies ensure almost complete transmission through the foil).

Protons reasonably simulate ballistic effects when considering particle size and mass [10,11]. Nevertheless, transmutation products such as helium with a critical contribution to damage evolution [12] cannot be addressed in proton irradiation studies. Self-ion irradiation as a surrogate for neutron irradiation was proved as feasible in various studies [11,13–17] focused on void swelling, but has been almost exclusively limited to TEM characterisation of regions a few microns thick.

Most irradiation studies are conducted at fluences up to ~10^{17} at/cm^2 [18] and in case of higher MeV energies these numbers are usually a few orders less. To be noted, however, these studies aimed at the investigation of basic ion-matter interaction and defect evolution. There are, however, few experiments with fluences surpassing that, such as the 5.42×10^{19} at/cm^2 proton irradiation performed recently [19]. Yet, to study engineering-relevant properties, bulk properties, the whole volume is to be irradiated homogeneously to introduce a quasi-uniform damage in terms of dpa, as well as in the case of evaluating the effect of transmutation helium, a "box-profile" of He concentration.

The ion beam centre at ATRI MTF STU recently upgraded its ion source systems to serve high-fluence high energy ion irradiations comprising a high-current upgrade of the

HVEE 358 Duoplasmatron ion source and the installation of a NEC TORVIS (Toroidal Volume Ion Source).

The aim is to perform multi-step ion-irradiation with different ion energies to achieve a nearly flat dpa-profile and almost constant irradiation hardening over the irradiated layer. The results of nanoindentation performed on such specimens will be easier to interpret as the substrate layer will not play a role in the results. Moreover, multi-step irradiation will increase the layer thickness compared with single step irradiation, approximately by a factor of 100. In this way, irradiation depths of ~70 μm can be achieved in steels, which allows applying micromechanical tests such as micropillar compression. This thickness of the radiation damage layer is not only sufficient for some micro-mechanical testing methods, but also for the application of conventional (radioisotope-based) positron sources in the techniques of positron annihilation spectroscopy (PAS), which is one of the important characterisation methods used in the post-irradiation examination (PIE).

2. Materials and Methods

2.1. Equipment

The 6 MV tandem accelerator setup of the Ion beam laboratory at ATRI MTF STU [20] has recently undergone upgrades, increasing the beam currents for proton and helium ions as well as the provision of end stations for performing high-fluence irradiation experiments. The new setup, Figure 1, has an upgraded HVEE 358 Duoplasmatron ion source with modified extraction optics and a new Na Charge-exchange canal (CEC) designed to deliver He^- beam currents up to 8 μA. The second addition is a used NEC TORVIS [21] with an Rb-CEC designed to deliver He^- and proton beams up to 20 μA and 100 μA respectively. These ion sources are complemented by a HVEE 860 Cs sputtering ion source for heavy ion beams. The related vacuum system is completely oil-free with a base vacuum level better than 5.0×10^{-7} mbar.

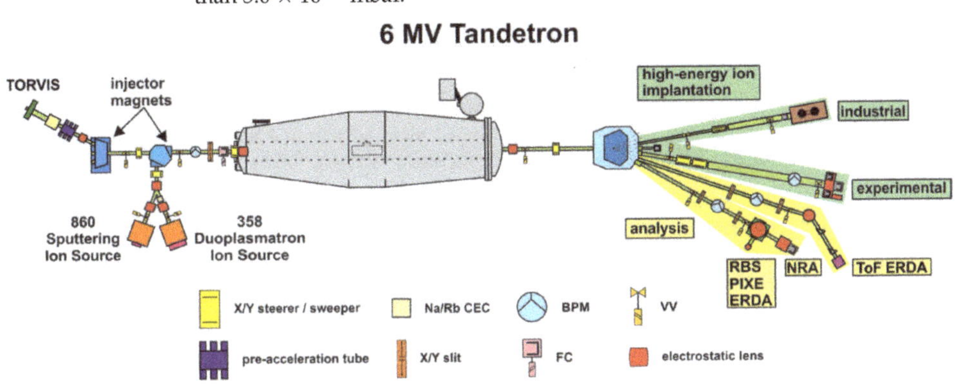

Figure 1. Schematic of the 6 MV Tandetron tandem accelerator setup.

On the high-energy side, the system is equipped with end-stations for analysis and high energy ion implantation/irradiation, Figure 2. The analytical end-station is equipped with standard Rutherford Backscattering Spectrometry (RBS), Particle Induced X-ray Emission (PIXE), and basic Elastic Recoil Detection Analysis (ERDA) for hydrogen and Nuclear Reaction Analysis (NRA). A detailed description of the analytical system can be found elsewhere [22]. The second analytical end-station, currently under procurement, will be equipped with a Time-of-Flight ERDA (ToF ERDA) spectrometer based on the Jyväskylä design [23] and will enable highly sensitive elemental composition analysis without the need for reference materials.

High-energy ion implantation/irradiation is served by two end-stations, a commercial semiconductor wafer handling system (client property), and another one for experimental purposes. The latter enables ion implantation/irradiation of substrates with sizes up to

Ø100 mm at room temperature and sample cooling down to LN2 temperatures. Sample heating up to 1000 °C is possible with Ø40 mm sample holding space. The usual experimental setup for high-fluence experiments is a water-cooled Ø40 mm sample holder where the beam is rastered over an area of 36 cm^2, which is given by the sample holding space and related current measurement system. With this setup, the ion currents abovementioned translate to a helium flux of 1.0×10^{12} at/cm^2/s and proton flux up to 1.2×10^{13} at/cm^2/s.

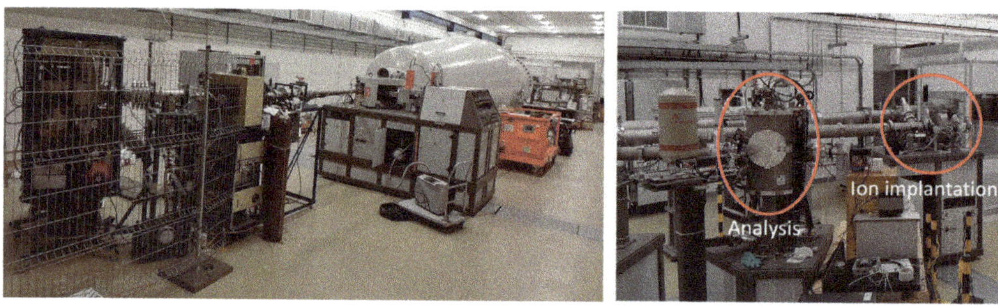

Figure 2. The 6 MV Tandetron accelerator with ion sources (**left**) and end-stations (**right**).

2.2. Multi Energy Sequential Irradiation Experiment Design

Irradiation experiments aiming to investigate the effect of transmutation helium require a homogeneous He concentration distribution. We adopted a similar approach as that in ref. [24], wherein the authors "assembled" dopant box-profiles from a sequence of ion implantation steps with decreasing ion energies. First, the implantation depth profiles at individual ion energies were calculated using SRIM [25] and fitted using a suitable function. The ion ranges were Gaussian-like by nature and bi-Gaussian functions yielded the best fitting, Figure 3. Since the investigated materials within the ongoing research projects are mostly nuclear grade ferritic/martensitic steels, all SRIM calculations of range as well as displacement damage profiles used ^{56}Fe as the target material. The number of energies to be employed is a parameter of choice and has to be chosen reasonably with respect to the system's switching and re-tuning time constants.

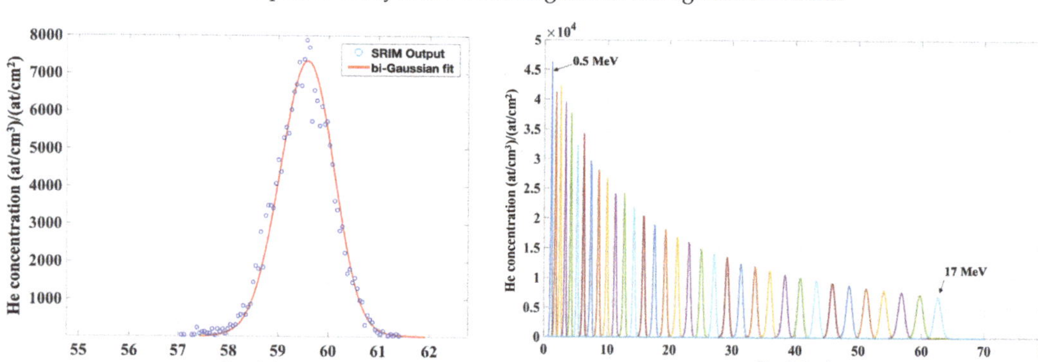

Figure 3. Fitting of individual concentration profiles (**left**) and the resulting profile-set (**right**).

The ion fluence at individual energies was calculated by solving the following minimisation problem:

$$\min_{u_{1\ldots n}} \left\| \sum_{i=1}^{n} C_{\mathrm{He}i}(x)u_i - W(x) \right\| \text{ subject to } \sum_{i=1}^{n} u_i = \int W(x)dx, \quad (1)$$

where $C_{Hei}(x)$ is the i-th helium concentration profile, $W(x)$ is the desired concentration profile (in our case a "box-profile"), u_i are ion fluences at individual energies, n is the number of implantation steps/energies. The resultant implantation profile becomes

$$\hat{W}(x) = \sum_{i=1}^{n} C_{Hei}(x)\hat{u}_i \tag{2}$$

where $\hat{u}_i$ are calculated optimal fluences at individual energies.

In light of the above, we decided to assemble the final profile of 34 individual ion implantations starting at 17 MeV going down to 500 keV in 500 keV steps, Figure 3. Our experiment aimed at reaching minimum 1000 appm (atomic ppm) He concentration, which yields bubble sizes suitable for TEM observation, as our previous research showed [12]. This concentration corresponds to a fluence of 5.42 × 10^{17} at/cm^2 and was limited by the achievable ion-beam current and acceptable duration, i.e., cost, of the experiment. The total irradiation time is on the order of a few hundreds of hours and was performed using the water-cooled sample holder kept at room temperature to avoid temperature effects. The resulting displacement damage across the irradiated region was calculated to 0.162 dpa, according to suggestions and recommendations published in ref. [26] using the NRT model [27]. It is important to note that the resulting He/dpa ratio is approximately 50× higher than the typical irradiation conditions of the spallation neutron targets (~100 appm He/dpa). Despite this, the helium concentration is almost two orders of magnitude higher than the expected helium production in fusion tokamaks; the planned microstructural characterisation and micromechanical testing can provide valuable experimental data to the material research for both fusion and spallation environments.

3. Results and Discussion

3.1. Upgraded Equipment

The upgraded HVEE 358 duoplasmatron ion source routinely operates with a 3–7 µA He^- injection current. Ion transport efficiency through the accelerator is still to be improved, as it is roughly about 50% just due to losses in the Ar stripper channel. The NEC TORVIS system was tested with hydrogen as well as helium, where we achieved stable proton currents around 30 µA, and 4 µA for helium. Much more is expected; however, these values were achieved during the first runs after the revival of the TORVIS and further tuning will increase ion yield.

3.2. High-Fluence Helium Irradiation

The multi-energy ion implantation experiment yields a 65 µm-thick irradiated layer with 1000 appm implanted helium, approximately homogeneously distributed in the layer, Figure 4. This makes micromechanical testing by micropillar compression as well as microcantilever bending possible in reasonable pillar and cantilever sizes to extract engineering-relevant data.

Accumulated irradiation damage was 0.162 dpa and was distributed in accordance with the irradiation profile, Figure 5. In comparison with using degrader foils [28], this approach enables better control of the helium concentration as well as damage profiles.

When considering irradiation fluxes, one has to keep in mind that aside from sample heating, which in our case was mitigated by a water-cooled copper/aluminium sample holder, 17 MeV He irradiation of Fe–Cr-based alloys produces a significant amount of neutron and gamma radiation. Our measurements indicated neutron dose rates up to 4.5 mSv/h/µA and gamma dose rates up to 0.5 mSv/h/µA at 1 m distance from the irradiation spot. The beam was He^{2+}, and hence 1 µA represents 3.12 × 10^{12} alpha particles per second. After the test phase, the chamber was additionally shielded with 5 cm lead shielding to protect the equipment and electronics present in the laboratory. Nevertheless, the laboratory is shielded by 1.5 m thick high-density concrete (3.8 g/cm^3) shielding, which attenuates the radiation down to background levels at the outer walls even at maximum beam currents [29].

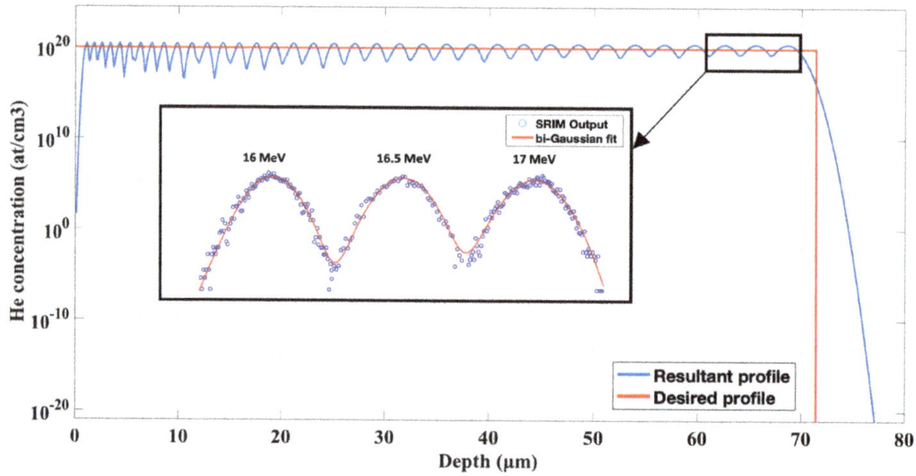

Figure 4. The resulting helium concentration profile assembled of 34 implantation steps.

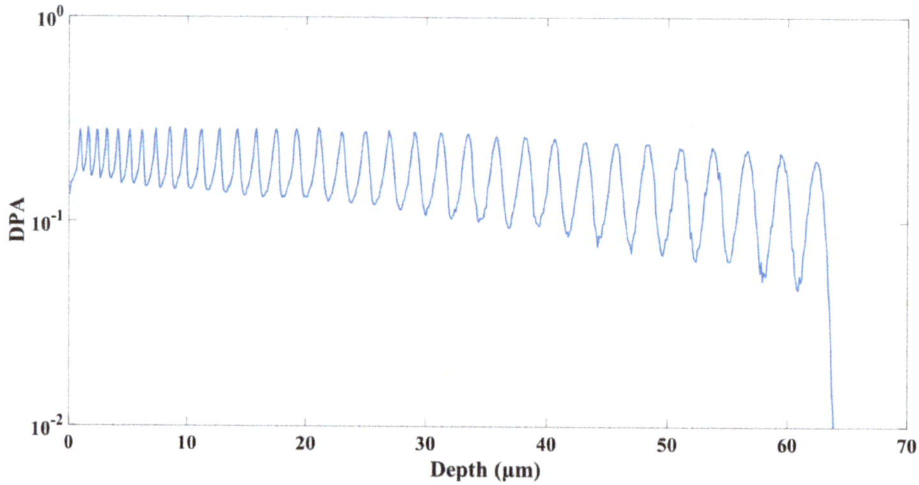

Figure 5. Resultant damage profile after 34 implantation steps.

4. Conclusions

The ATRI MTF STU ion beam laboratory upgraded its 6 MV tandem accelerator setup. High-current ion sources, the upgraded HVEE 358 duoplasmatron, and the NEC TORVIS increased current output especially for helium, enabling unique irradiation studies of radiation effects in fusion or spallation structural materials. The first tests were performed, and after further tuning the system is expected to deliver 10 μA He and 50–100 μA proton beams in the experimental chamber. Maximum sample size with heating up to 1000 °C or water cooling is Ø40 mm, otherwise up to Ø100 mm. The laboratory operates in open-access mode. The planned multi-energy high-fluence irradiation experiment will provide a 65 μm-thick approximately homogeneously irradiated layer in steels, enabling micromechanical testing and the evaluation of engineering-relevant properties of the irradiated materials.

Author Contributions: Conceptualization, P.N. and V.K.; methodology P.N., V.K. and M.K.; validation M.K.; investigation P.N., M.K., F.F. and Z.S.; resources J.D. and R.R.; writing—original draft preparation P.N. and Z.S.; writing—review & editing Z.S. and V.K.; project administration, P.N. and R.R.; funding acquisition, P.N. and R.R. All authors have read and agreed to the published version of the manuscript.

Funding: The authors would like to acknowledge partial support from the Slovak Research and Development Agency grants No. APVV-20-0010, APVV-20-0220, VEGA grant No. 1/0692/22 as well as from the European Regional Development Fund, projects No. ITMS2014+: 313011W085 and ITMS2014+: 313011BUH7.

Institutional Review Board Statement: Not applicable.

Informed Consent Statement: Not applicable.

Acknowledgments: Thanks to Roman Böttger, HZDR Innovation GmbH., for providing the NEC TORVIS and related ion transport system. The engagement and technical support of the ATRI accelerator technicians, Dušan Vaňa, Martin Muška, Juraj Halanda and Matúš Beňo is acknowledged as well.

Conflicts of Interest: The authors declare no conflict of interest.

References

1. DOENE. *A Technology Roadmap for Generation IV Nuclear Energy Systems*; DOENE (USDOE Office of Nuclear Energy, Science and Technology (NE)): Washington, DC, USA, 2002. [CrossRef]
2. Stanculescu, A. GIF R&D Outlook for Generation IV Nuclear Energy Systems: 2018 Update. In Proceedings of the Generation IV International Forum, Paris, France, 16–18 October 2018; p. 96. Available online: https://www.gen-4.org/gif/jcms/c_108744/gif-r-d-outlook-for-generation-iv-nuclear-energy-systems-2018-update (accessed on 16 August 2022).
3. Zinkle, S.J.; Snead, L.L. Designing Radiation Resistance in Materials for Fusion Energy. *Annu. Rev. Mater. Res.* **2014**, *44*, 241–267. [CrossRef]
4. Wheldon, C. Applications in Nuclear Physics and Nuclear Industry. In Proceedings of the PSD12: The 12th International Conference on Position Sensitive Detectors, Birmingham, UK, 12–17 September 2021; Available online: https://indico.cern.ch/event/797047/contributions/3638201/ (accessed on 16 August 2022).
5. University of Birmingham. High Flux Accelerator-Driven Neutron Facility. Available online: https://www.birmingham.ac.uk/research/activity/nuclear/about-us/facilities/high-flux-neutron-facility.aspx (accessed on 16 August 2022).
6. Bernardi, D.; Arbeiter, F.; Cappelli, M.; Fischer, U.; García, A.; Heidinger, R.; Krolas, W.; Martín-Fuertes, F.; Miccichè, G.; Muñoz, A.; et al. Towards the EU fusion-oriented neutron source: The preliminary engineering design of IFMIF-DONES. *Fusion Eng. Des.* **2019**, *146*, 261–268. [CrossRef]
7. ESFRI. *ESFRI Roadmap 2021 Strategy Report on Research Infrastructures*; Departmento di Fisica—Università degli Studi di Milano: Milan, Italy, 2021; p. 172. ISBN 978-88-943243-6-5. Available online: https://roadmap2021.esfri.eu/media/1295/esfri-roadmap-2021.pdf (accessed on 16 August 2022).
8. Jepeal, S.J.; Danagoulian, A.; Kesler, L.A.; Korsun, D.A.; Lee, H.Y.; Schwartz, N.; Sorbom, B.N.; Velez Lopez, E.; Hartwig, Z.S. An accelerator facility for intermediate energy proton irradiation and testing of nuclear materials. *Nucl. Instrum. Meth. Phys. Res. B* **2021**, *489*, 41–49. [CrossRef]
9. Hosemann, P. Small-scale mechanical testing on nuclear materials: Bridging the experimental length-scale gap. *Scripta Mater.* **2018**, *143*, 161–168. [CrossRef]
10. Rayaprolu, R.; Möller, S.; Linsmeier, C.; Spellerberg, S. Simulation of neutron irradiation damage in tungsten using higher energy protons. *Nucl. Mater. Energy* **2016**, *9*, 29–35. [CrossRef]
11. Was, G.S. Challenges to the use of ion irradiation for emulating reactor irradiation. *J. Mater. Res.* **2015**, *30*, 1158–1182. [CrossRef]
12. Krsjak, V.; Shen, T.; Degmova, J.; Sojak, S.; Korpas, E.; Noga, P.; Egger, W.; Li, B.; Slugen, V.; Garner, F.A. On the helium bubble swelling in nano-oxide dispersion-strengthened steels. *J. Mater. Sci. Technol.* **2022**, *105*, 172–181. [CrossRef]
13. Garner, F.A. Impact of the injected interstitial on the correlation of charged particle and neutron-induced radiation damage. *J. Nucl. Mater.* **1983**, *117*, 177–197. [CrossRef]
14. Was, G.S.; Jiao, Z.; Getto, E.; Sun, K.; Monterros, A.M.; Maloy, S.A.; Anderoglu, O.; Sencer, B.H.; Hackett, M. Emulation of reactor irradiation damage using ion beams. *Scr. Mater.* **2014**, *88*, 33–36. [CrossRef]
15. Gigax, J.G.; Kim, H.; Chen, T.; Garner, F.A.; Shao, L. Radiation instability of equal channel angular extruded T91 at ultra-high damage levels. *Acta Mater.* **2017**, *132*, 395–404. [CrossRef]
16. Shao, L.; Wei, C.-C.; Gigax, J.; Aitkaliyeva, A.; Chen, D.; Sencer, B.H.; Garner, F.A. Effect of defect imbalance on void swelling distributions produced in pure iron irradiated with 3.5 MeV self-ions. *J. Nucl. Mater.* **2014**, *453*, 176–181. [CrossRef]
17. Kumar, N.A.P.K.; Li, C.; Leonard, K.J.; Bei, H.; Zinkle, S.J. Microstructural stability and mechanical behavior of FeNiMnCr high entropy alloy under ion irradiation. *Acta Mater.* **2016**, *113*, 230–244. [CrossRef]

18. Krsjak, V.; Degmova, J.; Noga, P.; Petriska, M.; Sojak, S.; Saro, M.; Neuhold, I.; Slugen, V. Application of positron annihilation spectroscopy in accelerator-based irradiation experiments. *Materials* **2021**, *14*, 6238. [CrossRef] [PubMed]
19. Shiau, C.-H.; Sun, C.; McMurtrey, M.; O'Brien, R.; Garner, F.A.; Shao, L. Orientation-selected micro-pillar compression of additively manufactured 316L stainless steels: Comparison of as-manufactured, annealed, and proton-irradiated variants. *J. Nucl. Mater.* **2022**, *566*, 153739. [CrossRef]
20. Noga, P.; Dobrovodský, J.; Vaňa, D.; Beňo, M.; Závacká, A.; Muška, M.; Halgaš, R.; Minárik, S.; Riedlmajer, R. A new ion-beam laboratory for materials research at the Slovak University of Technolog. *Nucl. Insturm. Meth. Phys. Res. B* **2017**, *409*, 264–267. [CrossRef]
21. Hauser, T.M.; Daniel, R.E.; Norton, G.A.; Schroeder, J.B. High current He$^-$ injector for tandem accelerators. *Nucl. Instrum. Meth. Phys. Res. B* **2006**, *249*, 932–934. [CrossRef]
22. Dobrovodský, J.; Beňo, M.; Vaňa, D.; Bezák, P.; Noga, P. The first year operation experience with Ion Beam Analysis at the new STU Ion Beam Laboratory. *Nucl. Insturm. Meth. Phys. Res. B* **2019**, *450*, 168–172. [CrossRef]
23. Laitinen, M.; Rossi, M.; Julin, J.; Sajavaara, T. Time-of-flight—Energy spectrometer for elemental depth profiling—Jyväskylä design. *Nucl. Insturm. Meth. Phys. Res. B* **2014**, *337*, 55–61. [CrossRef]
24. Wu, H.; Böttger, R.; Couffignal, F.; Gutzmer, J.; Krause, J.; Munnik, F.; Renno, A.D.; Hübner, R.; Wiedenbeck, M.; Ziegenrücker, R. 'Box-Profile' Ion Implants as Geochemical Reference Materials for Electron Probe Microanalysis and Secondary Ion Mass Spectrometry. *Geostand. Geoanal. Res.* **2019**, *43*, 531–541. [CrossRef]
25. Ziegler, J.F.; Ziegler, M.D.; Biersack, J.P. SRIM—The stopping and range of ions in matter. *Nucl. Insturm. Meth. Phys. Res. B* **2010**, *268*, 1818–1823. [CrossRef]
26. Stoller, R.E.; Toloczko, M.B.; Was, G.S.; Certain, A.G.; Dwaraknath, S.; Garner, F.A. On the use of SRIM for computing radiation damage exposure. *Nucl. Insturm. Meth. Phys. Res. B* **2013**, *310*, 75–80. [CrossRef]
27. Norgett, M.J.; Robinson, M.T.; Torrens, I.M. A proposed method of calculating displacement dose rates. *Nucl. Eng. Des.* **1975**, *33*, 50–54. [CrossRef]
28. Brimbal, D.; Meslin, E.; Henry, J.; Décamps, B.; Barbu, A. He and Cr effects on radiation damage formation in ion-irradiated pure iron and Fe-5.40 wt.% Cr: A transmission electron microscopy study. *Acta Mater.* **2013**, *61*, 4757–4764. [CrossRef]
29. *Radiation Safety Report on Radiation Shielding Test of CAMBO Ion Beam Centre Building*; Slovak University of Technology: Bratislava, Bratislava, 2015.

materials

MDPI

Article

A Novel BCC-Structure Zr-Nb-Ti Medium-Entropy Alloys (MEAs) with Excellent Structure and Irradiation Resistance

Zhenqian Su [1,2], Zhaodong Quan [1,2], Tielong Shen [3,4,*], Peng Jin [3,4], Jing Li [3,4], Shiwen Hu [1,2] and Dexue Liu [1,2,*]

1 School of Materials Science and Engineering, Lanzhou University of Technology, Lanzhou 730050, China
2 State Key Laboratory of Advanced Processing and Recycling of Non-Ferrous Metals, Lanzhou University of Technology, Lanzhou 730050, China
3 Institute of Modern Physics, Chinese Academy of Sciences, Lanzhou 730050, China
4 University of Chinese Academy of Sciences, Beijing 100049, China
* Correspondence: shentielong@impcas.ac.cn (T.S.); dxliu@lut.edu.cn (D.L.); Tel.: +86-13609383013 (T.S.); +86-13919947009 (D.L.); Fax: +86-931-4969639 (T.S.); +86-931-2976702 (D.L.)

Abstract: Medium-entropy alloys (MEAs) are prospective structural materials for emerging advanced nuclear systems because of their outstanding mechanical properties and irradiation resistance. In this study, the microstructure and mechanical properties of three new single-phase body-centered cubic (BCC) structured MEAs ($Zr_{40}Nb_{35}Ti_{25}$, $Zr_{50}Nb_{35}Ti_{15}$, and $Zr_{60}Nb_{35}Ti_{5}$) before and after irradiation were investigated. It is shown that the yield strength and elongation after fracture at room temperature are greater than 900 MPa and 10%, respectively. Three MEAs were irradiated with 3 MeV Fe^{11+} ions to 8×10^{15} and 2.5×10^{16} ions/cm^2 at temperatures of 300 and 500 °C, to investigate the irradiation-induced hardening and microstructure changes. Compared with most conventional alloys, the three MEAs showed only negligible irradiation hardening and even softening in some cases. After irradiation, they exhibit somewhat surprising lattice constant reduction, and the microstructure contains small dislocation loops. Neither cavities nor precipitates were observed. This indicates that the MEAs have better irradiation resistance than traditional alloys, which can be attributed to the high-entropy and lattice distortion effect of MEAs.

Keywords: Zr-Nb-Ti MEAs; microstructural characterization; irradiation resistance; defects evolution

Citation: Su, Z.; Quan, Z.; Shen, T.; Jin, P.; Li, J.; Hu, S.; Liu, D. A Novel BCC-Structure Zr-Nb-Ti Medium-Entropy Alloys (MEAs) with Excellent Structure and Irradiation Resistance. *Materials* **2022**, *15*, 6565. https://doi.org/10.3390/ma15196565

Academic Editor: Adam Grajcar

Received: 27 July 2022
Accepted: 7 September 2022
Published: 22 September 2022

1. Introduction

Rapidly increasing energy demands and climate change concerns inevitably emphasize the role of clean energy in the future while making fossil fuel infrastructure obsolete. After thermal power and hydropower, nuclear energy has grown to become the world's third-largest energy source [1]. Nowadays, the GIF (Generation Nuclear Energy International Forum) has already proposed the Generation IV Nuclear System, promising competitiveness and advancement in safety, economy, sustainable development, non-proliferation, etc. However, the structure materials for the reactor will withstand very harsh environments in terms of high-temperature, high-dose irradiation, and corrosion [2], posing a significant challenge to its mechanical properties and dimensional changes attribute to void swelling or creep [1]. Hence, in order to develop advanced nuclear reactor systems, research on high-performance materials with outstanding irradiation resistance and corrosion resistance is critical. Since the 1950s, the light water reactor (LWR) has been the leading type of reactor for electricity production. Because Zr-based alloys have a low thermal neutron absorption cross-section, outstanding thermal conductivity and superior corrosion resistance, they are used as major materials for fuel cladding materials in this reactor [3,4]. At temperatures exceeding 1200 °C, however, Zr-based alloys accelerate oxidation and hydrogen pick-up, causing significant embrittlement [5–7], such as in the Fukushima Daiichi accident in 2011. As a result, advanced reactor systems require an upgraded Zr-based alloy or alternative materials with improved accident tolerance [8].

Recently, HEAs have attracted great attention for their excellent mechanical properties [9,10], corrosion resistance [11], and radiation resistance [12–15]. Typically, HEAs are composed of five or more elements in equal or near-equal atomic ratios [16,17], such as different variants of FeCoNiCrMn alloy, also known as Cantor alloy. The outstanding performance of HEAs is attributed to some typical characteristics, including the high-entropy effect, the severe-lattice-distortion effect, the sluggish diffusion effect, and the cocktail effect. High-stability HEAs thus may be promising candidate materials for advance nuclear systems.

The effect of the atomic number, type, and grouping of alloying components on irradiation (by ions, electrons, and neutrons)-prompted microstructural evolution and property deterioration has received much interest recently. The defect development process in alloys with high chemical complexity is slowed down by improving energy dissipation as well as defect formation and migration energy, according to recent findings [18]. Although HEA requires five or more elements, it is gradually being demonstrated that high entropy effects do not necessarily dominate the alloy structure and that the number of elements does not determine performance. HEAs have fewer densities of vacancy and interstitial defect clusters owing to their enhanced vacancy-interstitial recombination in the cascade compared to traditional alloys [19–21]. However, the accumulation of damage in several HEAs with varying chemical complexity has been studied experimentally. For instance, Granberg et al. likewise found a significant decrease in damage gathering under delayed irradiation in NiFe and NiCoCr compared to elemental Ni, which has been demonstrated by TEM images and MD simulations [22]. Olsen et al. additionally found defect cluster sizes yet with higher densities in NiCo contrasted to Ni [23]. Yang et al. investigated the FCC CrMnFeCoNi HEA demonstrating a very stable structure when compared to 304ss and pure Ni, both have a remarkable helium-ion irradiation resilience at room temperature and 450 °C [24]. Zhang et al. firstly discovered the dislocation loops in BCC HEAs after helium-ion irradiation at 350 °C [25].

When designing HEAs to meet the requirements of the nuclear system and the formation of the BCC structure, elements with a low thermal neutron absorption cross-section and excellent mechanical properties are considered. In addition to the Zr, there is a small amount of Nb in Zr_4 alloy which has been used as the cladding material in pressurized water reactors. Further, the addition of Ti can enhance the corrosion performance of the alloy. In view of this, novel Zr-Nb-Ti MEAs with BCC structure was designed. To replicate neutron irradiation, the MEAs were subjected to heavy-ion irradiation. The effects of irradiation on crystal structure, hardness, and microstructure evolution were studied.

2. Experimental Procedures

2.1. Samples Preparation

The three alloy ingots of $Zr_{40}Nb_{35}Ti_{25}$, $Zr_{50}Nb_{35}Ti_{15}$, and $Zr_{60}Nb_{35}Ti_5$, nominal compositions of Zr, Nb, and Ti with a purity of >99.95 wt.%, were carefully weighted, followed by melting using a cold crucible levitation technique under argon atmosphere. Additionally, they were flipped and remelted 4–5 times to ensure their chemical homogeneity, which was checked by an FEI Quanta 450 SEM (Scanning Electron Microscope) equipped with EDX (energy dispersive spectrometer). Finally, it was processed into the diverse size and shape for subsequent testing using a cutting machine. After being processed, the cross-sections of the samples were mounted in resin for the surface treatment. The grinding of the surface with sizes of 120# to 2000# and polishing to the mirror-finish surface was achieved using Tripoli, intermediates, and a finishing rough. The nominal compositions of the three HEAs are listed in Table 1. For convenience, the three HEAs were designated as 1#, 2#, and 3#, respectively.

Table 1. Nominal composition (at. %) of the three HEAs.

Sample No.	HEAs	Zr	Nb	Ti
1#	$Zr_{40}Nb_{35}Ti_{25}$	45.06	40.16	14.78
2#	$Zr_{50}Nb_{35}Ti_{15}$	53.46	38.12	8.42
3#	$Zr_{60}Nb_{35}Ti_{5}$	61.05	36.28	2.67

2.2. Heavy Ion Irradiation Procedures

The center parts of the as-cast ingots were chopped into specimens with the dimension of $10 \times 10 \times 1$ mm^3. They were mechanically polished using 1.5 and 0.5 μm diamond pastes after being ground with SiC abrasive paper from 200 to 3000 grit. Finally, specimens were polished for over 8 h with 40 nm colloidal silica slurry to remove the residual strain caused by grinding and mechanical polishing.

The irradiation experiments were carried out at the Chinese Academy of Science's Institute of Modern Physics 320 kV platform for multi-discipline research with highly charged ions. The three alloys were irradiated with 3 MeV Fe^{11+} (defocused beam) at 300 and 500 °C to ion fluences of 8×10^{15} ions/cm^2 and 2.5×10^{16} ions/cm^2. The effect of irradiation condition on the microstructure of 1#, 2#, and 3# specimens can be investigated by comparison with each other with low and high fluences and low temperature and high temperature. To make ion irradiation doses consistent with hypothetical neutron irradiations, using the SRIM simulation with lattice and surface binding energies of the Zr, Nb, and Ti atoms set to zero, and we used the Kinchin–Pease calculation mode based on findings from previous studies [26]. Figure 1 shows the depth profile of damage and Fe ion concentration calculated by the SRIM 2008. For the displacement energies of the lattice atoms Zr, Nb, and Ti, the value is 40, 30, and 20.8 eV, respectively. Table 2 shows the results for the major factors, including fluence, temperature, peak damage, and peak dose rate.

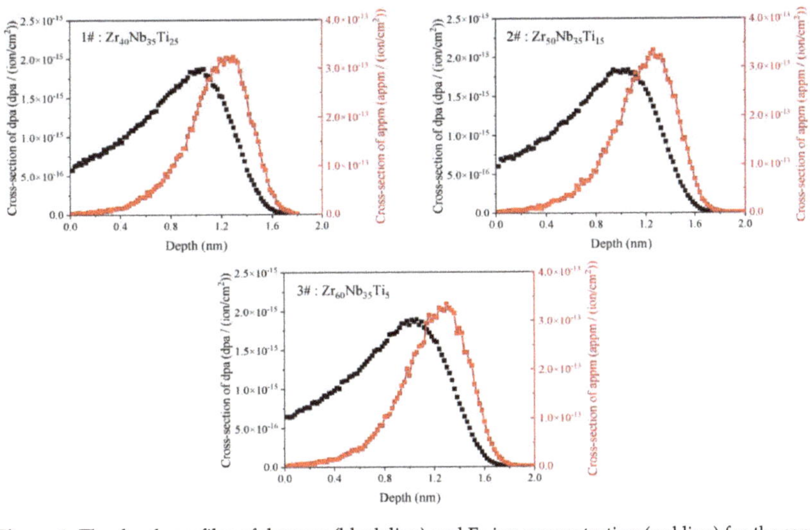

Figure 1. The depth profiles of damage (black line) and Fe ion concentration (red line) for the samples **1#**, **2#**, and **3#** calculated by SRIM 2008.

Table 2. The irradiation conditions of 3 MeV Fe^{11+} for the three HEAs, including temperature, fluence, and peak displacement damage, and peak dose rate calculated by SRIM 2018.

Samples	Temperature (°C)	Irradiation Fluence (ions/cm^2)	Peak Damage (dpa)	Peak Dose Rate (dpa/s)
1#/2#/3#	300	8×10^{15}	15	~6.5×10^{-4}
		2.5×10^{16}	47	
	500	8×10^{15}	15	
		2.5×10^{16}	47	

2.3. *Depth-Sensing Nanoindentation Methods*

Hardness measurements on samples were performed at room temperature utilizing an Anton Paar nanoindenter with Berkovich diamond indenter (3-sided pyramidal tip) for test. All the tests were performed in the continuous stiffness measurement mode (CSM) [27], with a consistent stacking rate set to 0.15 nm/s and a surface methodology speed for the tip of 10 nm/s. The entrance profundity analyzed in the examples went from 0 to 1500 nm, with a greatest heap of 200 mN, and the information inside 200 nm from the example surface was disposed of because of enormous information disperses related to surface unpleasantness. For good measurable examination, each example was indented with approximately 6 to 8 indents of each example, and the normal of the outcomes was utilized in the examination. Since the actual measured nanohardness value is easily influenced by the depth of the indenter diffusion, often the actual measured depth is 4–5-fold deeper than the depth at that point, we selected the measured value at a depth of 400 nm as the actual nanohardness value based on our experience and to fit the application of the Nix–Gao model [28]. Thus, the hardness at an indenter profundity of 400 nm was picked as the ideal condition for estimating just the 3 MeV Fe^{11+} irradiation zone without the significant information disperse seen at shallower profundities.

2.4. *Characterization and Properties Procedure*

The phase structures of original alloys were detected by x-ray diffraction (XRD) using a D/max-2400 diffractometer with Cu Ka1 radiation. The XRD parameters are voltage 45 kV, current 50 mA, Cu Kα target, X-ray wavelength 0.15406 nm, scanning range 20~100°, and scanning step 5 °/min at room temperature. However, the damage layer of irradiated samples is shallow, and the conventional X-rays penetrate more deeply than the region. This leads to the obtained information coming from both the damaged and undamaged regions. Therefore, the changes in the composition and crystalline phase were investigated by using grazing incidence X-ray diffraction spectroscopy (GIXRD). Usually, the penetration depth can be calculated using the total reflection critical Angle model [29]. The microstructures in HEAs were analyzed using transmission electron microscopy (TEM). The TEM lamellae were prepared by dual-beam focused-ion beam (FIB) lift-out method, and the microstructure was observed under the bright-field image model by FEI Tecnai F20 TEM operated at 200 kV. The room temperature mechanical properties and high-temperature mechanical properties were evaluated using Instron 5565 universal testing machine (INSTRON, Norwood, MA, USA) at the strain rate of 0.01 s^{-1}. The samples utilized in the tensile test were cut into canine bone shape by wire, the gauge length and cross-sectional area of the specimens were 8 mm and (1 × 2) mm^2, separately, and the tensile fracture surfaces were observed by SEM.

3. Results and Discussion

3.1. *Microstructure and Mechanical Behavior of the Unirradiated Zr-Nb-Ti MEAs*

Figure 2 shows the XRD patterns and optical images of these three MEAs. It can be seen from Figure 2a that Zr-Nb-Tix MEAs ($Zr_{40}Nb_{35}Ti_{25}$, $Zr_{50}Nb_{35}Ti_{15}$, and $Zr_{60}Nb_{35}Ti_5$) have five diffraction peaks of BCC structure, including (110), (200), (211), (220) and (310). The strongest peaks are located on the (110) crystal plane and no other diffraction peaks appear, which indicates that the three alloys had a single BCC structure with a lattice constant of 0.3351, 0.3395, and 0.3422 nm, respectively. The diffraction peaks shift gradually towards the smaller 2θ values, which indicates that the lattice constant increases with Zr content due to the atomic radius of Zr being larger than that of Nb and Ti, resulting in lattice expansion. Using empirical formula to calculate the lattice constant of these three alloys is 0.3410, 0.3430, and 0.3448 nm, respectively, which is consistent with the values of the experimental. The SEM images of the as-cast samples reveal a homogeneous dendrites microstructure.

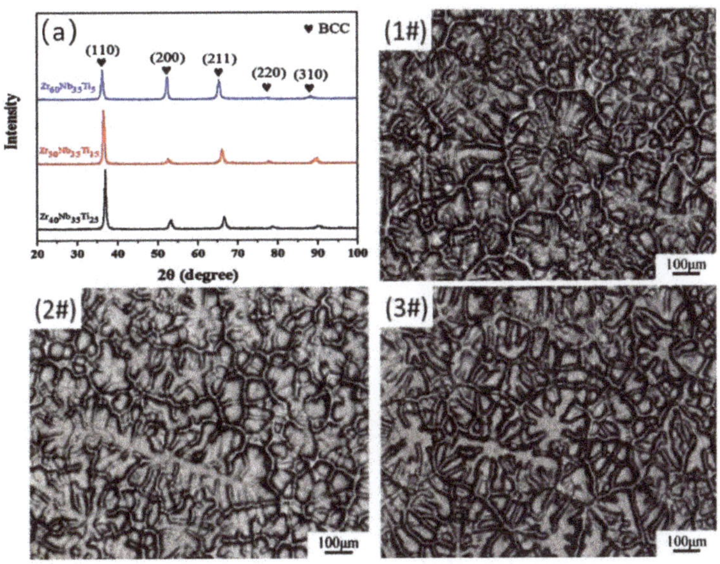

Figure 2. (**a**) XRD patterns and optical images of $Zr_{40}Nb_{35}Ti_{25}$ (**1#**), $Zr_{50}Nb_{35}Ti_{15}$ (**2#**), and $Zr_{60}Nb_{35}Ti_{5}$ (**3#**) origin alloys.

Several metrics for predicting the structure stability and phase formation of HEAs have been developed thus far [30]. Zhang et al. suggested a criterion for the formation of solid-solution phases in HEAs based on the enthalpy of mixing (ΔH_{mix}) and atomic radius difference (δ) [31], as well as an extra parameter was also proposed as $\Omega = T_m \Delta S_{mix} / |\Delta H_{mix}|$, where T_m is the average melting temperature and ΔS_{mix} the entropy of mixing of an alloy. From available data, the conditions for the creation of HEAs with a single solid-solution phase are $\Omega \geq 1.1$ and $\delta \leqq 6.6\%$, respectively. According to the electronic structure theory, Guo et al. [32] introduced another measure named valence electron concentration (VEC) to predict the phase stability of HEAs. According to Guo's statistics, BCC solid-solution phases are stable since the VEC value is less than 6.87. Table 3 shows the specifications of these MEAs, which are perfectly in accordance with the aforementioned criteria.

Table 3. Calculation of relevant parameters of the $Zr_{40}Nb_{35}Ti_{25}$, $Zr_{50}Nb_{35}Ti_{15}$, and $Zr_{60}Nb_{35}Ti_{5}$ high-entropy alloys.

Alloys	**ΔH_{mix} (KJ/mol)**	**ΔS_{mix} (J/mol·K)**	**Ω**	**δ (%)**	**VEC**	**Δχ**	**T_m (K)**
$Zr_{40}Nb_{35}Ti_{25}$	2.94	8.98	7.009	4.18	4.35	0.122	2294.74
$Zr_{50}Nb_{35}Ti_{25}$	3.22	8.302	5.98	4.23	4.35	0.127	2317.8
$Zr_{60}Nb_{35}Ti_{5}$	3.5	6.848	4.57	4.11	4.35	0.129	2336.5

Figure 3 shows the tensile true stress–strain curve of three HEAs at room temperature and 400 °C. Table 4 summarizes the three HEAs' yield strength (σ_y), ultimate strength (δ_u), and elongation after fracture ($\varepsilon_{ef, \%}$). The yield strength of these three MEAs at room temperature is 945 MPa, 903 MPa, and 1028 MPa, and the strain is 17.5%, 14%, and 11%. The yield strength of these three MEAs at 400 °C is 380–400 MPa at 400 °C, the tensile strength is 390–430 MPa, and the elongation rate exceeds 20%. Notably, these three MEAs also have shown excellent mechanical properties at 400 °C, even the fracture strength was reduced by about 40%, and the elongation increased by about 1 time to 28%. The fracture strength of these three MEAs is considerably higher than that of most other FCC structures HEAs, which may indicate a single BCC phase with strong bonding inherited from the refractory elements and the high solution hardening effect of the whole-solute matrix. Meanwhile, these MEAs exhibit significant plastic strain, which is uncommon for BCC HEAs.

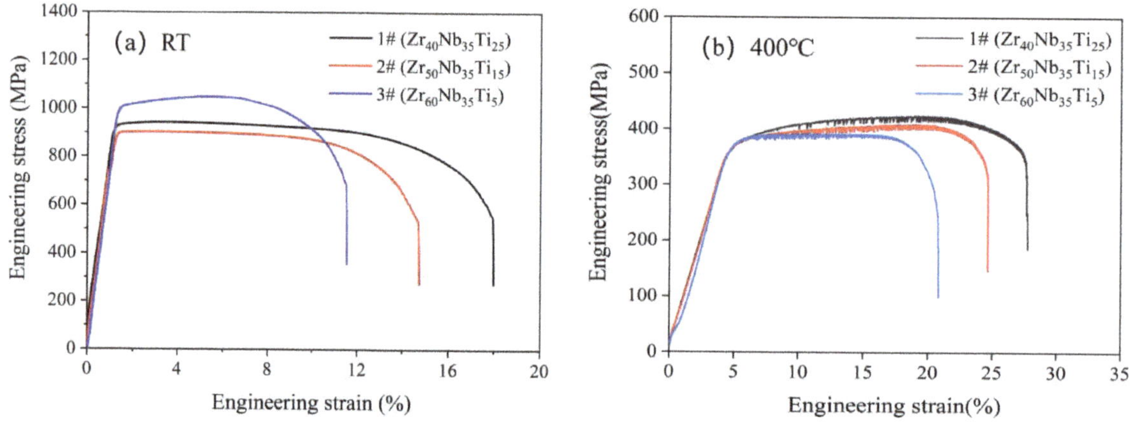

Figure 3. Engineering stress–strain curves of the alloys under tensile at (**a**) room temperature and (**b**) 400 °C.

Table 4. The yield strength (σ_y), ultimate tensile (σ_u), and elongation after fracture (ε_{ef}) for the ZrNbTi MEAs.

Alloys	Temperature (°C)	Yield Strength (δ_y, MPa)	Ultimate Tensile (δ_u, MPa)	Elongation after Fracture ($\varepsilon_{ef, \%}$)
$Zr_{40}Nb_{35}Ti_{25}$ (1#)		945	1010	17.5
$Zr_{50}Nb_{35}Ti_{15}$ (2#)	Room	903	915	14
$Zr_{60}Nb_{35}Ti_{5}$ (3#)		1028	1050	11
$Zr_{40}Nb_{35}Ti_{25}$ (1#)		350	423	28
$Zr_{50}Nb_{35}Ti_{15}$ (2#)	400	355	408	25
$Zr_{60}Nb_{35}Ti_{5}$ (3#)		360	390	20.5

Figure 4 shows the SEM images of the fracture surface for these three HEAs. Obvious necking near the fracture surface can be found in Figure 4a, which demonstrates that typical plastic deformation took place before fracture. As can be observed in the SEM picture, the grain boundaries are severely deformed and reveal clear vein patterns inside the crystal grains, which are distributed as densely spaced dimples on the fracture with diameters between 50 and 100 μm. Some tearing was seen beyond grain boundaries, and the tearing ridges were heavily surrounded by small dimples less than 5 μm in size. This indicates that the alloys are all more resistant to fracture and shear deformation. All of these properties, as shown in Figure 4, indicate that the fracture behavior is that of a ductile fracture, which is consistent with the significant plastic strain.

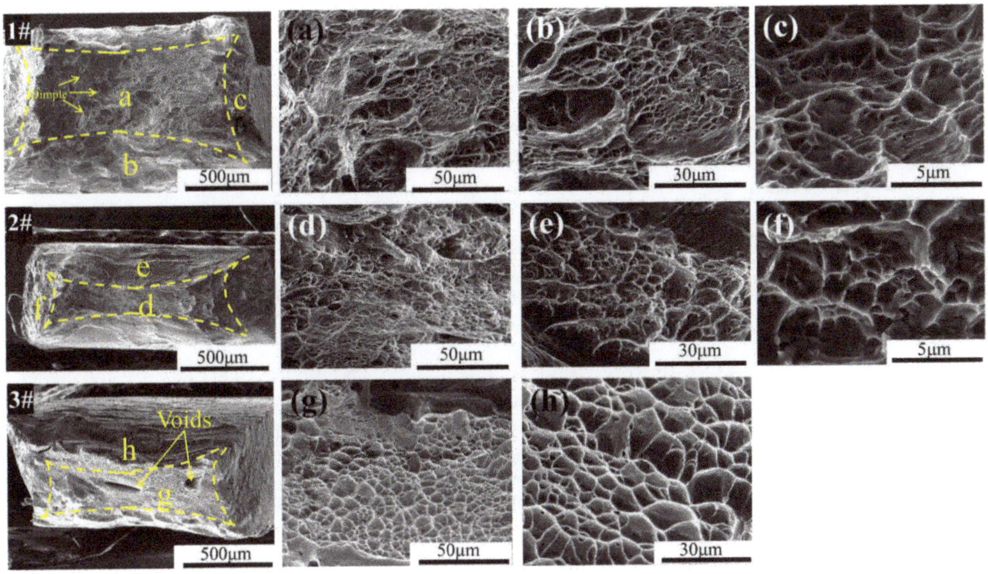

Figure 4. SEM images of fracture morphology for samples $Zr_{40}Nb_{35}Ti_{25}$ (**1#**), $Zr_{50}Nb_{35}Ti_{15}$ (**2#**), and $Zr_{60}Nb_{35}Ti_{5}$ (**3#**) after tensile tests. (**a–c**) are enlargements of the zones labeled "a", "b", and "c" in the sample 1#, respectively. Correspondingly, (**d–f**) are the local enlargements of the sample 2#, and (**g**,**h**) is the local enlargements of the sample 3#.

3.2. XRD Analysis of the Irradiated MEAs

Grazing incidence X-ray diffraction spectroscopy (GIXRD) tests were performed to investigate the irradiation effect on the crystal structure of these three MEAs, which are shown in Figure 5. Figure 5a shows $Zr_{40}Nb_{35}Ti_{25}$ MEAs exhibit BCC structure with (110) and (200) planes preferred orientation. After irradiation to a fluence of 8×10^{15} and 2.5×10^{16} ions/cm^2 and a temperature of 300 and 500 °C, the $Zr_{40}Nb_{35}Ti_{25}$ MEA remains stable BCC structure without a secondary phase. However, the intensity of BCC diffraction peaks with (200) and (211) planes decreases drastically and cannot be simply connected to the 8×10^{15} ions/cm^2 irradiation at 500 °C. Figure 5b shows $Zr_{50}Nb_{35}Ti_{15}$ MEAs exhibit BCC structure with (110) and (200) planes preferred orientation. After irradiation to a fluence of 8×10^{15} and 2.5×10^{16} ions/cm^2 and a temperature of 300 and 500 °C, the $Zr_{50}Nb_{35}Ti_{15}$ MEA still remains as a stable BCC structure without a secondary phase. Figure 5c shows $Zr_{60}Nb_{35}Ti_{5}$ MEAs exhibit BCC structure with (110) and (211) planes preferred orientation. After irradiation to a fluence of 8×10^{15} and 2.5×10^{16} ions/cm^2 and a temperature of 300 and 500 °C, the $Zr_{40}Nb_{35}Ti_{25}$ MEA also remains stable BCC structure without a secondary phase. However, the intensity of BCC diffraction peaks with the (200) plane decrease drastically and even disappears, and peaks with the (211) plane increase seriously after the 8×10^{15} ions/cm^2 irradiation at 500 °C. In terms of diffraction peaks decreasing dramatically or even disappearing, Sun et al. [33] observed that the intensity of bcc diffraction peaks decreased dramatically, or even can hardly be distinguished after helium-ion irradiation for the $Al_{1.5}CrFeNi$ HEA film. Moreover, Zhang et al. [25] studied the influence of ion irradiation on the two BCC MoNbCrVTi and MoNbCrZrTi HEAs, observing a dramatic decrease in the intensity of XRD peaks. This was attributed to the amorphous structure forming as a result of high-level stress. The decreases in diffraction peak seen in this study can be attributed to local melting and recrystallization of amorphous materials caused by thermal spikes created by irradiation [20].

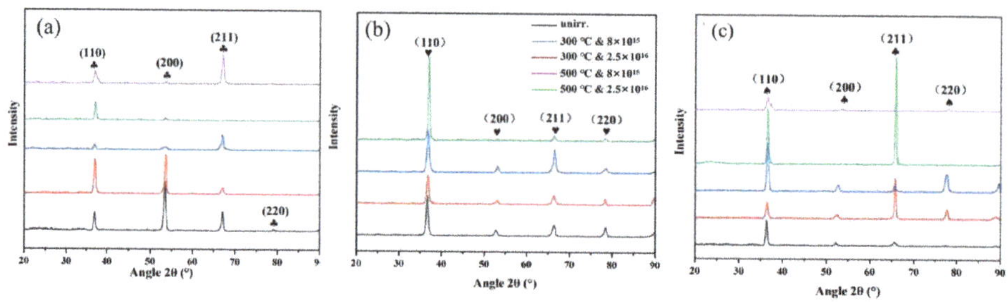

Figure 5. XRD patterns of the unirradiated and irradiated MEAs at different ion influences and temperatures. (**a**) $Zr_{40}Nb_{35}Ti_{25}$ (**1#**), (**b**) $Zr_{50}Nb_{35}Ti_{15}$ (**2#**), and (**c**) $Zr_{60}Nb_{35}Ti_{5}$ (**3#**).

The diffraction peaks of these three MEAs shift gradually towards the bigger 2θ values after irradiation, implying a decrease in lattice constant. As shown in Figure 5, the lattice constant of these three MEAs has varying degrees of reduction compared to the original material within the different irradiated conditions. The observed irradiation-caused shrink of lattice constant in our MEAs is different from the conventional alloys, such as 316 L, 304 H, and Zr-Nb alloys, of which the lattice constant expands after irradiation [34,35]. Lu et al. [36] detected a drop in the $Ti_2ZrHfV_{0.5}Mo_{0.2}$ HEA's lattice constant with an irradiation dose of 3×10^{16} ions/cm^2. On the other hand, Zhang et al. [14] discovered the irradiation response of the CoCrCuFeNi HEA, which had a rising lattice constant with an average dose of 0.73–368.5 dpa. In this work, the decrease in the lattice constant may be ascribed to the irradiation relaxes the extreme lattice distortion that exists in the HEAs due to different atomic sizes of the solutes, eventually resulting in the lattice shrink.

3.3. The Nanoindentation Result of the Irradiated MEAs

To further study the performance variation of these three MEAs after irradiation, the nanoindentation hardness of unirradiated and irradiated MEAs samples was measured. Nanoindentation is a valuable technology for monitoring mechanical property changes in the small affected regions. To confirm the correctness of the experimental results, six single indents were created for each sample in this investigation. As shown in Figure 6, the average nanoindentation hardness of all the studied samples was plotted as a function of the indentation depth.

The hardness decreases along with the depth in both irradiated and un-irradiated samples due to the indentation-size effect, that can also lead to the hardness value increasing with decreasing indentation size. In our work, the specimens were prepared with vibration polishing, which produced a smooth and deformation-free surface with minimal surface irregularity. Hence, the data within 100 nm are not shown since the near-surface hardness data exhibit a large scatter due to the surface irregularities [37], and indentations depth was controlled to the same depth at 1900 nm for all materials to remove the size effect [38]. It was demonstrated that the largest region of the near-surface ion irradiation region, which is independent of the underlying unirradiated substrate, corresponds to an indenter depth of 350 nm [28]. This is due to the fact that the elastic stress fields beneath the indenter are sensitive to microstructure features that are up to 10-fold the indenter depth. In order to conduct a quantitative analysis of the nanoindentation hardness of the ion irradiation specimens, indenter depths ranging from 200 to 400 nm were utilized. Figure 6a shows that the irradiated $Zr_{40}Nb_{35}Ti_{25}$ MEA sample has a larger hardness than the unirradiated sample with the increase the irradiation fluence and temperature. Figure 6b shows that the irradiated $Zr_{50}Nb_{35}Ti_{15}$ MEA sample's hardness also increases, except for the sample irradiated with 8×10^{15} ions/cm^2 at 300 °C. Instead, Figure 6c shows that the $Zr_{60}Nb_{35}Ti_{5}$ MEA has a smaller hardness than the unirradiated, and the irradiated for 8×10^{15} ions/cm^2 at 500 °C. has no obvious variation. The magnitude error bar is attributed to scattering and difficulty in fitting the data shown in Figure 6. Figure 7 illustrates the Nix–Gao model for three MEAs. The result is the same as its indentation depth profiles.

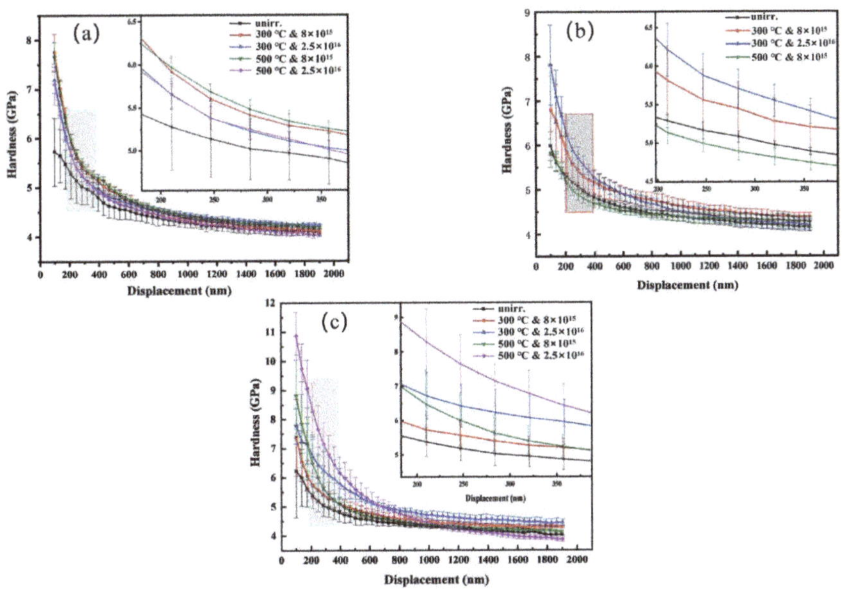

Figure 6. Variation of hardness with indentation depth before and after Fe^{11+} irradiation at different ion influences and temperatures: (**a**) $Zr_{40}Nb_{35}Ti_{25}$, (**b**) $Zr_{50}Nb_{35}Ti_{15}$, and (**c**) $Zr_{60}Nb_{35}Ti_5$.

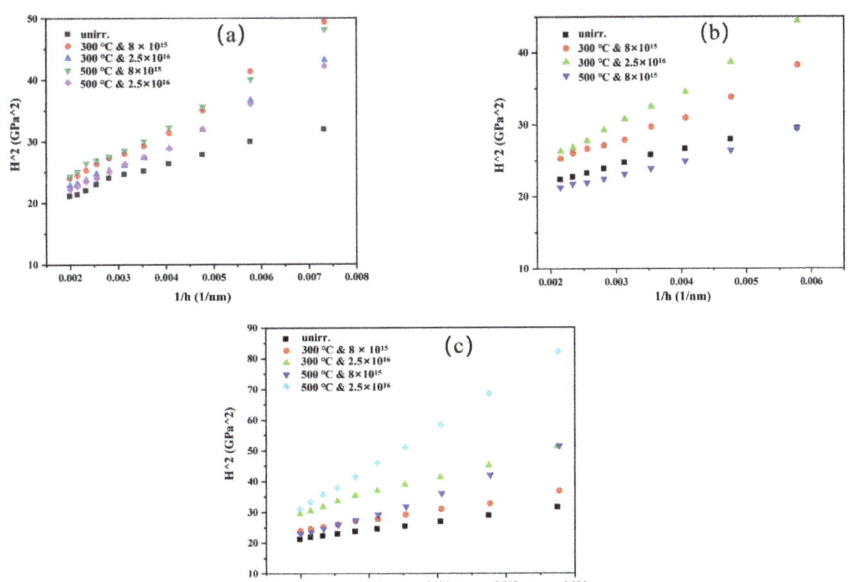

Figure 7. Nix–Gao model for three MEAs before and after Fe^{11+} irradiation: (**a**) $Zr_{40}Nb_{35}Ti_{25}$, (**b**) $Zr_{50}Nb_{35}Ti_{15}$, and (**c**) $Zr_{60}Nb_{35}Ti_5$.

Table 5 illustrates the results of irradiation hardening and the hardening rate on the alloy. The results show the hardness value after irradiation with a high dose and high temperature are to varying changes, some materials increased, and others decreased after irradiation, but the overall pattern is still hardening. Several factors could cause irradiation-hardening, including the radiation-induced precipitation (RIP) and the formation of dislocation loops [39]. Additionally, it has been reported that the visible dislocation loops are the main contributor to the irradiation hardening of HEAs at room temperature [40,41]. As far as we know, this is firstly showing a softening of the BCC single-solution MEAs after irradiation.

Table 5. Average nanohardness and hardening rate of Zr-Nb-Ti MEAs in the range of 200–400nm from the surface under different irradiation conditions.

Alloys	$Zr_{40}Nb_{35}Ti_{25}$ (GPa)	$Zr_{50}Nb_{35}Ti_{15}$ (GPa)	$Zr_{60}Nb_{35}Ti_{5}$ (GPa)	Hardening Rate (%)		
				$Zr_{40}Nb_{35}Ti_{25}$	$Zr_{50}Nb_{35}Ti_{15}$	$Zr_{60}Nb_{35}Ti_{5}$
Unirradiated	5.17	5.15	5.22	——	——	——
$8 \times 10^{15}/300$ °C	5.56	5.53	5.62	7.54	7.38	7.66
$2.5 \times 10^{16}/300$ °C	5.48	5.84	6.42	6.01	13.40	22.99
$8 \times 10^{15}/500$ °C	5.66	4.95	5.9	9.48	−3.88	13.03
$2.5\times 10^{16}/500$ °C	5.43	——	7.5	5.03	——	42.72

Irradiation would cause hardening, embrittlement, and even softening of structure materials, which is among the main failure modes of structural materials in nuclear power plants. The three MEAs have better hardness stability than most typical alloys, which could be due to fewer flaws caused by irradiation in MEAs.

3.4. Irradiation Defects of the MEAs

TEM examinations of the samples were carried out to further confirm the structural development of the samples brought on by the irradiation. The initial Zr-Nb-Ti MEAs' TEM pictures revealed a very dense microstructure. Only a tiny number of imperfections linked to the ion milling process were found. After exposure to radiation, the cross-sectional microstructures of Zr-Nb-Ti MEAs samples are shown in bright-field (BF) TEM images in Figure 8. Figure 8a shown the sample was irradiated at 500 °C to a fluence of 8×10^{15} ions/cm^2, which equates to a peak damage dose of 15 dpa, as shown in the TEM image Figure 8a. The photos clearly show the differences between the irradiation-induced faults and the reference (unirradiated) samples. In contrast to the SRIM predictions, there are more black dots in the peak damage region and the defect bands of the $Zr_{40}Nb_{35}Ti_{25}$ samples are concentrated in the depth range between 450 and 700 nm. In contrast to $Zr_{40}Nb_{35}Ti_{25}$ samples, which are mostly scattered at a depth of 500–800 nm, the irradiation-induced defects of $Zr_{50}Nb_{35}Ti_{15}$ and $Zr_{60}Nb_{35}Ti_{5}$ samples demonstrate higher-density dispersion. Materials with flaws brought on by irradiation typically have voids, dislocation loops, and chemical segregation. The dislocation loop is often the most frequent defect in materials that have been exposed to radiation, and we were unable to detect any voids or precipitates in the TEM pictures. Additionally, samples with various Zr concentrations, as well as those exposed to various irradiation effects and temperatures, exhibit the irradiation-induced defects in a variety of microstructural ways. Therefore, in the materials with various Zr concentrations, we described and examined the irradiation-induced dislocation loop. It is believed that the increase in Zr content increases the high entropy effect and lattice distortion effect of the three MEAs, resulting in a larger lattice distortion degree of the alloy and some point defects annihilation, reducing the damage degree of the alloy. The outcomes of the MEAs with various Zr contents that were exposed to various radiation circumstances may be explained by looking at the nucleation and development stages of dislocation loops. More Zr contents increased the lattice distortion effect, which is used to pin irradiation-induced the defects produced during irradiation and inhibit their migration or accumulation, according to the perspective of the defect distribution. Because of this, the doped samples had a greater defect density and smaller defect size than the MEAs samples with less Zr content.

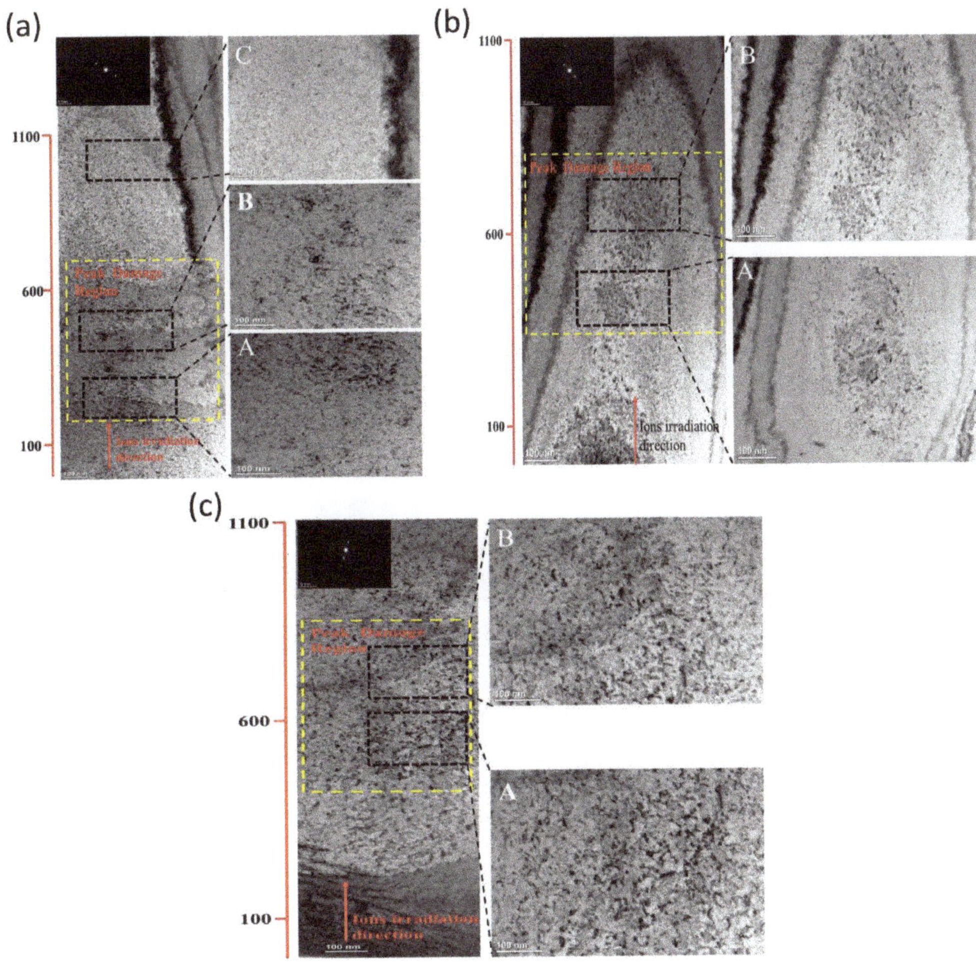

Figure 8. BF-TEM images of irradiated samples showing the irradiation-induced black-dot defects (**a**) 500 °C—8 × 10^{15} sample of $Zr_{40}Nb_{35}Ti_{25}$, (**b**) 500 °C—8 × 10^{15} sample of $Zr_{50}Nb_{35}Ti_{15}$, and (**c**) 500 °C—2.5 × 10^{16} sample of $Zr_{60}Nb_{35}Ti_5$. The g = <110> near the [111] zone axis was selected. A, B, and C are the local enlargements of each sample along the irradiation direction.

Figure 9 shows the morphology and size of the defect of $Zr_{40}Nb_{35}Ti_{25}$ samples with different irradiated conditions. Obviously, the defects in $Zr_{40}Nb_{35}Ti_{25}$ samples at 300 °C to a fluence of 8 × 10^{15} ions/cm^2 are mainly black dots that are consist of the small dislocation loops and defect clusters. In the samples irradiated at 300 °C to a fluence of 2.5 × 10^{16} ions/cm^2 that can see more black dots with an increment of the fluence. The microstructure of another sample irradiated at 500 °C to a fluence of 2.5 × 10^{16} ions/cm^2 contains mainly large-sized dislocation loops and dislocation lines. The aggregation of self-interstitial atoms or clusters causes these dislocation loops, which are mostly interstitial in nature. Irradiation with 3 MeV Fe^{11+} ions triggered collision cascades and the formation of interstitial atoms. These interstitial atoms clump together to create defect clusters, which absorb more interstitials or other point-defect clusters, eventually producing dislocation loops and lines. With increasing irradiation impact and temperature, there is a modest rise in loop size and a drop in loop density.

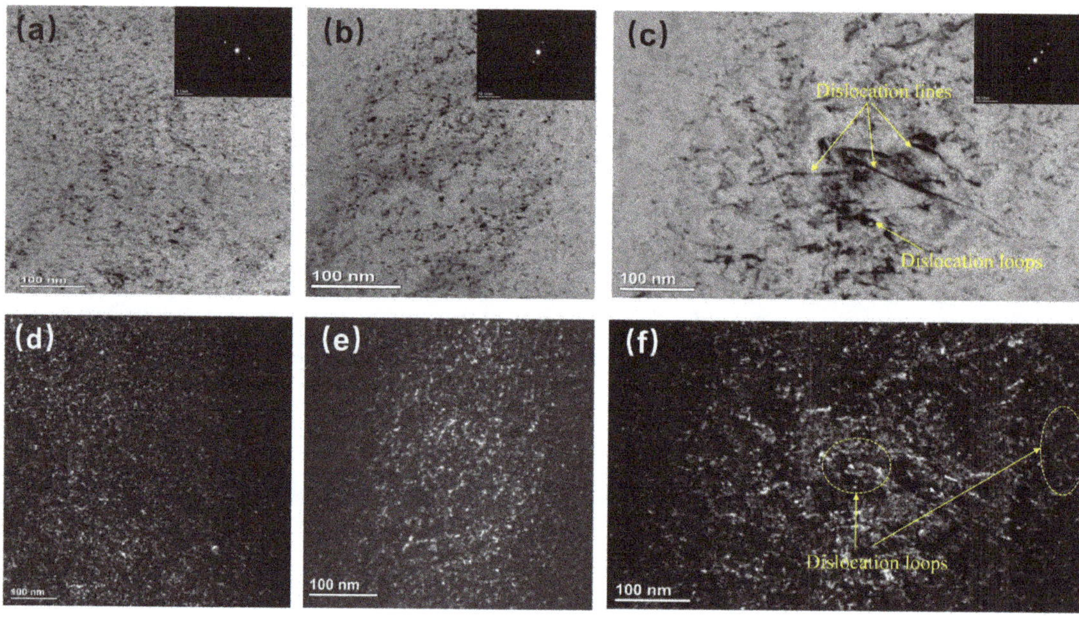

Figure 9. Enlarged BF and DF TEM images of $Zr_{40}Nb_{35}Ti_{25}$ samples irradiated by 3 MeV Fe^{11+} ions: (**a**) 500 °C—8 × 10^{15}, (**b**) 300 °C—2.5 × 10^{16}, and (**c**) 500 °C—2.5 × 10^{16}. (**d–f**) The corresponding weak-beam DF images. The g = <110> near the [111] zone axis was selected.

The irradiated samples did not include any holes or radiation-induced segregation (RIS). Materials exposed to radiation at doses greater than 1 dpa frequently include voids because irradiation-induced vacancies consolidate into stable cavities. Dimensional expansion and a reduction in fracture toughness are two effects of voids. It is believed that radiation-induced segregation, which occurs when alloying elements segregate out of equilibrium near sinks, is a key cause of irradiation-assisted stress corrosion cracking. As a result of the severe lattice distortion in the Three MEAs, irradiation-induced defects may diffuse slowly, inhibiting the growth of defect clusters, and defect annihilation may take place during the transition from amorphization to crystallization under conditions of high atomic-level stress.

4. Conclusions

In this work, we designed and prepared single BCC Zr-Nb-Ti MEAs, and then performed the Fe-ion irradiation experiments to examine their irradiation resistance. The nanoindentation experiment and microstructural characterization indicated excellent properties of the fabricated materials in terms of irradiation resistance compared with traditional alloys. The main conclusions are summarized as follows:

(1) The tensile tests showed that the MEAs have a good combination of strength and toughness. At room temperature, the yield strength of these three alloys, $Zr_{40}Nb_{35}Ti_{25}$, $Zr_{50}Nb_{35}Ti_{15}$, and $Zr_{60}Nb_{35}Ti_{5}$, is 943, 903, and 1285 MPa, and the fracture strain is 17.5%, 14%, and 11%, respectively. At 400 °C, the fracture strain increases to 28%, 25%,

and 20.5%, respectively. The fracture morphology shows that the fracture mode is the ductile fracture.

(2) The nanoindentation test showed the Zr-Nb-Ti MEAs have a little irradiation hardening that increased with the irradiation fluence.

(3) After irradiation, contrary to traditional alloys, the XRD diffraction peaks of Zr-Nb-Ti MEAs were shifted to the right, indicating a decrease in the lattice constant. No visible phase transformation or decomposition of Zr-Nb-Ti MEAs was observed.

(4) Only dislocation loops and dislocation lines were observed in the peak damage region. This suggests the irradiation resistance of the MEAs is better than that of traditional alloys.

Author Contributions: Conceptualization, Z.S., T.S. and D.L.; methodology, Z.S., Z.Q., S.H., P.J. and J.L.; formal analysis, Z.S., Z.Q. and S.H.; resources, T.S. and D.L.; data curation, Z.S., Z.Q., S.H., P.J. and J.L.; writing—original draft preparation, Z.S.; writing—review and editing, Z.S., P.J., T.S. and D.L.; project administration, T.S. and D.L.; funding acquisition, T.S. and D.L. All authors have read and agreed to the published version of the manuscript.

Funding: This research was funded by the National Natural Science Foundation of China (Grant Nos. 12162023, 51365029, 12175288 and U1832206) and the International Cooperation project of Gansu Province under grant number (Grant No. 20YF8WA064).

Institutional Review Board Statement: Not applicable.

Informed Consent Statement: Not applicable.

Data Availability Statement: Not applicable.

Conflicts of Interest: The authors declare no conflict of interest.

References

1. Zinkle, S.J.; Was, G.S. Materials challenges in nuclear energy. *Acta Mater.* **2013**, *61*, 735–758. [CrossRef]
2. Stork, D.; Zinkle, S.J. Introduction to the special issue on the technical status of materials for a fusion reactor. *Nucl. Fusion* **2017**, *57*, 092001. [CrossRef]
3. Charit, I. Accident Tolerant Nuclear Fuels and Cladding Materials. *Jom* **2017**, *70*, 173–175. [CrossRef]
4. Duan, Z.; Yang, H.; Satoh, Y.; Murakami, K.; Kano, S.; Zhao, Z.; Shen, J.; Abe, H. Current status of materials development of nuclear fuel cladding tubes for light water reactors. *Nucl. Eng. Des.* **2017**, *316*, 131–150. [CrossRef]
5. Terrani, K.A.; Zinkle, S.J.; Snead, L.L. Advanced oxidation-resistant iron-based alloys for LWR fuel cladding. *J. Nucl. Mater.* **2014**, *448*, 420–435. [CrossRef]
6. Zinkle, S.J.; Terrani, K.A.; Snead, L.L. Motivation for utilizing new high-performance advanced materials in nuclear energy systems. *Curr. Opin. Solid State Mater. Sci.* **2016**, *20*, 401–410. [CrossRef]
7. Motta, A.T.; Chen, L.-Q. Hydride Formation in Zirconium Alloys. *Jom* **2012**, *64*, 1403–1408. [CrossRef]
8. Zinkle, S.J.; Terrani, K.A.; Gehin, J.C.; Ott, L.J.; Snead, L.L. Accident tolerant fuels for LWRs: A perspective. *J. Nucl. Mater.* **2014**, *448*, 374–379. [CrossRef]
9. Gludovatz, B.; Hohenwarter, A.; Catoor, D.; Chang, E.H.; George, E.P.; Ritchie, R.O. A fracture-resistant high-entropyalloy for cryogenic applications. *Science* **2014**, *345*, 1153–1158. [CrossRef]
10. Rackwitz, J.; Yu, Q.; Yang, Y.; Laplanche, G.; George, E.P.; Minor, A.M.; Ritchie, R.O. Effects of cryogenic temperature and grain size on fatigue-crack propagation in the medium-entropy CrCoNi alloy. *Acta Mater.* **2020**, *200*, 351–365. [CrossRef]
11. Chen, Y.Y.; Duval, T.; Hung, U.D.; Yeh, J.W.; Shih, H.C. Microstructure and electrochemical properties of high entropy alloys—A comparison with type-304 stainless steel. *Corros. Sci.* **2005**, *47*, 2257–2279. [CrossRef]
12. Zhang, Y.; Jin, K.; Xue, H.; Lu, C.; Olsen, R.J.; Beland, L.K.; Ullah, M.W.; Zhao, S.; Bei, H.; Aidhy, D.S.; et al. Influence of chemical disorder on energy dissipation and defect evolution in advanced alloys. *J. Mater. Res.* **2016**, *31*, 2363–2375. [CrossRef]
13. Zhang, Y.; Stocks, G.M.; Jin, K.; Lu, C.; Bei, H.; Sales, B.C.; Wang, L.; Beland, L.K.; Stoller, R.E.; Samolyuk, G.D.; et al. Influence of chemical disorder on energy dissipation and defect evolution in concentrated solid solution alloys. *Nat. Commun.* **2015**, *6*, 8736. [CrossRef] [PubMed]
14. Zhang, Y.; Tunes, M.A.; Crespillo, M.L.; Zhang, F.; Boldman, W.L.; Rack, P.D.; Jiang, L.; Xu, C.; Greaves, G.; Donnelly, S.E.; et al. Thermal stability and irradiation response of nanocrystalline CoCrCuFeNi high-entropy alloy. *Nanotechnology* **2019**, *30*, 294004. [CrossRef]
15. Zhang, Y.; Zhao, S.; Weber, W.J.; Nordlund, K.; Granberg, F.; Djurabekova, F. Atomic-level heterogeneity and defect dynamics in concentrated solid-solution alloys. *Curr. Opin. Solid State Mater. Sci.* **2017**, *21*, 221–237. [CrossRef]
16. Cantor, B.; Chang, I.T.H.; Knight, P.; Vincent, A.J.B. Microstructural development in equiatomic multicomponent alloys. *Mater. Sci. Eng. A* **2004**, *375–377*, 213–218. [CrossRef]

17. Yeh, J.-W.; Lin, S.-K.C.S.-J.; Gan, J.-Y.; Chin, T.-S.; Shun, T.-T.; Tsau, C.-H.; Chang, S.-Y. Nanostructured High-Entropy Alloys with Multiple Principal Elements Novel Alloy Design Concepts and Outcomes. *Adv. Eng. Mater.* **2004**, *6*, 299–303. [CrossRef]
18. Zhao, S.; Zhang, Y.; Weber, W.J. High Entropy Alloys: Irradiation. In *Reference Module in Materials Science and Materials Engineering*; Elsevier: Amsterdam, The Netherlands, 2020.
19. Egami, T.; Ojha, M.; Khorgolkhuu, O.; Nicholson, D.M.; Stocks, G.M. Local Electronic Effects and Irradiation Resistance in High-Entropy Alloys. *Jom* **2015**, *67*, 2345–2349. [CrossRef]
20. Egami, T.; Guo, W.; Rack, P.D.; Nagase, T. Irradiation Resistance of Multicomponent Alloys. *Metall. Mater. Trans. A* **2013**, *45*, 180–183. [CrossRef]
21. Lu, C.; Niu, L.; Chen, N.; Jin, K.; Yang, T.; Xiu, P.; Zhang, Y.; Gao, F.; Bei, H.; Shi, S.; et al. Enhancing radiation tolerance by controlling defect mobility and migration pathways in multicomponent single-phase alloys. *Nat. Commun.* **2016**, *7*, 13564. [CrossRef]
22. Aidhy, D.S.; Lu, C.; Jin, K.; Bei, H.; Zhang, Y.; Wang, L.; Weber, W.J. Point defect evolution in Ni, NiFe and NiCr alloys from atomistic simulations and irradiation experiments. *Acta Mater.* **2015**, *99*, 69–76. [CrossRef]
23. Olsen, R.J.; Jin, K.; Lu, C.; Beland, L.K.; Wang, L.; Bei, H.; Specht, E.D.; Larson, B.C. Investigation of defect clusters in ion-irradiated Ni and NiCo using diffuse X-ray scattering and electron microscopy. *J. Nucl. Mater.* **2016**, *469*, 153–161. [CrossRef]
24. Yang, L.; Ge, H.; Zhang, J.; Xiong, T.; Jin, Q.; Zhou, Y.; Shao, X.; Zhang, B.; Zhu, Z.; Zheng, S.; et al. High He-ion irradiation resistance of CrMnFeCoNi high-entropy alloy revealed by comparison study with Ni and 304SS. *J. Mater. Sci. Technol.* **2019**, *35*, 300–305. [CrossRef]
25. Zhang, Z.; Han, E.-H.; Xiang, C. Irradiation behaviors of two novel single-phase bcc-structure high-entropy alloys for accident-tolerant fuel cladding. *J. Mater. Sci. Technol.* **2021**, *84*, 230–238. [CrossRef]
26. Stoller, R.E.; Toloczko, M.B.; Was, G.S.; Certain, A.G.; Dwaraknath, S.; Garner, F.A. On the use of SRIM for computing radiation damage exposure. *Nucl. Instrum. Methods Phys. Res. Sect. B Beam Interact. Mater. At.* **2013**, *310*, 75–80. [CrossRef]
27. Oliver, W.C. An improved technique for determining hardness and elastic modulus using load and displacement sensing indentation experiments. *J. Mater. Res.* **1992**, *7*, 1564–1583. [CrossRef]
28. Nix, W.D.; Gao, H. Indentation size effects in crystalline materials A law for strain gradient plasticity. *J. Mech. Phys. Solids* **1998**, *46*, 411–425. [CrossRef]
29. Lim, G.; Parrish, W.; Ortiz, C.; Bellotto, M.; Hart, M. Grazing incidence synchrotron x-ray diffraction method for analyzing thin films. *J. Mater. Res.* **1987**, *2*, 471–477. [CrossRef]
30. Zhang, Y.; Zhou, Y.J.; Lin, J.P.; Chen, G.L.; Liaw, P.K. Solid-Solution Phase Formation Rules for Multi-component Alloys. *Adv. Eng. Mater.* **2008**, *10*, 534–538. [CrossRef]
31. Yang, X.; Zhang, Y. Prediction of high-entropy stabilized solid-solution in multi-component alloys. *Mater. Chem. Phys.* **2012**, *132*, 233–238. [CrossRef]
32. Guo, S.; Liu, C.T. Phase stability in high entropy alloys: Formation of solid-solution phase or amorphous phase. *Prog. Nat. Sci. Mater. Int.* **2011**, *21*, 433–446. [CrossRef]
33. Sun, S.; Qiu, N.; Zhang, K.; He, P.; Ma, Y.; Gou, F.; Wang, Y. Segregation of $Al_{1.5}CrFeNi$ high entropy alloys induced by vacancy-type defects. *Scr. Mater.* **2019**, *161*, 40–43. [CrossRef]
34. Hosemann, P.; Frazer, D.; Fratoni, M.; Bolind, A.; Ashby, M.F. Materials selection for nuclear applications: Challenges and opportunities. *Scr. Mater.* **2018**, *143*, 181–187. [CrossRef]
35. Yvon, P.; Carré, F. Structural materials challenges for advanced reactor systems. *J. Nucl. Mater.* **2009**, *385*, 217–222. [CrossRef]
36. Lu, Y.; Huang, H.; Gao, X.; Ren, C.; Gao, J.; Zhang, H.; Zheng, S.; Jin, Q.; Zhao, Y.; Lu, C.; et al. A promising new class of irradiation tolerant materials: $Ti_2ZrHfV_{0.5}Mo_{0.2}$ high-entropy alloy. *J. Mater. Sci. Technol.* **2019**, *35*, 369–373. [CrossRef]
37. Odette, G.R.; Yamamoto, T.; Williams, T.J.; Nanstad, R.K.; English, C.A. On the history and status of reactor pressure vessel steel ductile to brittle transition temperature shift prediction models. *J. Nucl. Mater.* **2019**, *526*, 151863. [CrossRef]
38. Hosemann, P.; Kiener, D.; Wang, Y.; Maloy, S.A. Issues to consider using nano indentation on shallow ion beam irradiated materials. *J. Nucl. Mater.* **2012**, *425*, 136–139. [CrossRef]
39. Jin, K.; Lu, C.; Wang, L.M.; Qu, J.; Weber, W.J.; Zhang, Y.; Bei, H. Effects of compositional complexity on the ion-irradiation induced swelling and hardening in Ni-containing equiatomic alloys. *Scr. Mater.* **2016**, *119*, 65–70. [CrossRef]
40. Kumar, N.A.P.K.; Li, C.; Leonard, K.J.; Bei, H.; Zinkle, S.J. Microstructural stability and mechanical behavior of FeNiMnCr high entropy alloy under ion irradiation. *Acta Mater.* **2016**, *113*, 230–244. [CrossRef]
41. Zhang, Z.X.; Chen, D.S.; Han, W.T.; Kimura, A. Irradiation hardening in pure tungsten before and after recrystallization. *Fusion Eng. Des.* **2015**, *98–99*, 2103–2107. [CrossRef]

materials

MDPI

Review

Application of Positron Annihilation Spectroscopy in Accelerator-Based Irradiation Experiments

Vladimir Krsjak [1,2,*], Jarmila Degmova [1,2], Pavol Noga [2], Martin Petriska [1], Stanislav Sojak [1,2], Matus Saro [1], Igor Neuhold [1] and Vladimir Slugen [1]

1 Slovak University of Technology, Faculty of Electrical Engineering and Information Technology, Institute of Nuclear and Physical Engineering, Ilkovicova 3, 81219 Bratislava, Slovakia; jarmila.degmova@stuba.sk (J.D.); martin.petriska@stuba.sk (M.P.); stanislav.sojak@stuba.sk (S.S.); matus.saro@stuba.sk (M.S.); igor.neuhold@cern.ch (I.N.); vladimir.slugen@stuba.sk (V.S.)

2 Slovak University of Technology, Faculty of Materials Science and Technology, Advanced Technologies Research Institute, Jana Bottu 2781/25, 91724 Trnava, Slovakia; pavol.noga@stuba.sk

* Correspondence: vladimir.krsjak@stuba.sk

Abstract: Positron annihilation spectroscopy (PAS) is widely recognized as a powerful characterization technique in all types of radiation damage studies in nuclear materials. In the past, fission reactor irradiation of reactor pressure vessel (RPV) steels was a primary aim in most studies, while today's applications of PAS in this field are centered around ion implantation experiments in advanced structural materials. These experiments use hydrogen, helium, heavy ions, and their combination to simulate various radiation environments of future nuclear reactors or nuclear research facilities. The spectrum of ion energies used ranges from a few tens of keV to tens or even hundreds of MeV in proton irradiation or spallation neutron source irradiation experiments. The variety of ion energies, irradiation temperatures, and other experimental conditions poses a major challenge to researchers, who often fail to successfully incorporate the lessons learned from their research. In this paper, we review and supplement recent PAS studies in which structural materials irradiated under a variety of irradiation conditions were investigated using positron annihilation spectroscopy. It summarizes the most important conclusions and lessons learned from the application of PAS in accelerator-based irradiation experiments.

Keywords: ion irradiation; positron annihilation spectroscopy; positron lifetime; Doppler broadening spectroscopy; f/m steels; RPV steels; oxide dispersion strengthened steels

Citation: Krsjak, V.; Degmova, J.; Noga, P.; Petriska, M.; Sojak, S.; Saro, M.; Neuhold, I.; Slugen, V. Application of Positron Annihilation Spectroscopy in Accelerator-Based Irradiation Experiments. *Materials* **2021**, *14*, 6238. https://doi.org/10.3390/ma14216238

Academic Editor: Wen-Tong Geng

Received: 23 August 2021
Accepted: 12 October 2021
Published: 20 October 2021

1. Introduction

The experimental simulation of the harsh radiation environments of future nuclear fission and fusion reactors requires compromises to be made on the selection of bombarding particles and parameters such as flux, energy spectrum, or the production of transmutation elements and the change in the chemical composition of the target material. Additionally, for the characterization techniques employed, it is inevitable to consider the shape of the damage profile, particularly when using low to intermediate energy (tens of keV to hundreds of keV) of charged particles.

Among the numerous analytical techniques used in material irradiation studies [1–3], positron annihilation spectroscopy (PAS) is well known for its spectacular sensitivity to atomic-scale vacancy-type defects, and it has been widely used in the past [4]. This unique sensitivity originates from the fact that positrons are attracted to regions of the lattice with an open volume. The positron, unlike any other particle, acts as a self-seeking probe for vacancy-type defects in condensed matter. Due to its unique features and non-destructive nature, PAS has been recognized as a convenient complementary tool for the microstructural characterization of lattice defects.

The published data refer primarily to irradiation studies of reactor pressure vessel (RPV) steels [5,6]. Despite being a valuable contribution to understanding radiation-induced degradation processes in these materials, the lessons learned cannot be simply extrapolated to the new challenging radiation conditions of future nuclear fission and fusion reactors [7]. The use of fast neutron irradiation experiments to simulate the harsh radiation environments of future reactor systems faces two significant hurdles. While it provides reasonable displacement damage rates in producing collision cascades, it does not provide an adequate simulation of gaseous transmutation products, such as hydrogen or helium, severely impairing the defect recombination processes in the displaced matrix. The second, obvious complication arises from the induced activity of the neutron-irradiated samples; the handling of which requires dedicated nuclear facilities and radioisotope laboratories. Both issues can be conveniently solved by using ion implanters in either single-beam or dual-/triple-beam configurations. Various PAS studies of materials modified by ion bombardment, in a wide range of ion energies and fluences, have been published in the last decade. In addition to the structural materials addressed in this paper, plasma-facing materials [8–11] and nuclear fuel [12–14] have been investigated using this approach.

Comparably to ion irradiation experiments, slow positron beam experiments enable the region of interest within the studied sample to be precisely selected. This makes positron beam techniques natural approaches to studying accelerated radiation damage in materials. Most of the studies use slow positron beams with positron energy of up to a few tens of keV, corresponding to a mean implantation depth in the micrometer range in most nuclear-relevant materials. These studies have generally been aimed at ion implantation experiments using accelerating energies of up to a few MeV. There have been very few irradiation studies published on nuclear structural materials utilizing charged particle irradiation of the target at energies of 10–100 MeV [15–17]. An example of such an implantation (multi-energy He^+ implantation of structural steels) is discussed in the paper by Noga et al. in this Special Issue. Moving up with the particle accelerator energy, the irradiation experiments necessarily cause activation of the material; for instance, by spallation reactions [18]. This radiological complication is, however, outweighed by extremely interesting irradiation data, involving realistic and fusion-relevant displacement damage rates, helium and hydrogen production rates, as well as quasi-homogeneous damage distribution over a bulk sample [19]. Concerning PAS, the last feature mentioned enables the convenient use of radioisotope positron sources with a continuous energy spectrum to probe the irradiated material. The comparison between conventional (radioisotope-based) PAS techniques and slow positron beam techniques is discussed in the next chapter.

While the individual PAS techniques provide excellent reproducibility of the results in various studies of semiconductors and non-metallic solids, the results obtained from the investigations of realistic alloys exposed to dissimilar irradiation conditions can rarely be found to correlate between different research studies. Here, let us omit the fundamental distinction between the near-surface studies, utilizing slow positron beams, and the "bulk" studies, based on unmoderated radioisotope positron sources. Quite substantial discrepancies between the acquired results can arise from the same kind of post-irradiation PAS examination. There are several potential reasons for this, which can be divided into two categories. The first group of issues relates to the apparatus, signal processing, and data processing. In this regard, particular attention must be paid to suppressing the noise signal originating either from transmutation elements or Compton-scattered positron annihilation gamma. While three-detector positron lifetime spectrometers [20] and a coincidence setup of Doppler broadening spectrometers [21] can potentially solve this problem, the deviations in the geometry and activity of the used positron sources, as well as in the isotopic composition of the measured samples, affect the effectivity of this solution. Another source of discrepancy between different PAS characterizations, for instance, ion-implanted samples, comes from the lack of consensus in the data evaluation. While some studies report the peak values of displacement damage and ion concentration data, others report an integral value over the whole implantation profile, or an integral value over the probing particle stopping profile. It is important to note that the stopping profile of a monoenergetic positron is very broad at high-incident energies (tens of keV), and while the mean stopping depth of the positron can be accounted for the peak region of the ion-modified layer, a substantial amount of signal can come from either the substrate or the thermal spike region.

The first and foremost benefit of using PAS techniques in nuclear materials irradiation experiments is that it allows the detection of the slightest changes in the microstructure, associated either to defect production or defect recombination processes. Positron annihilation characteristics enable a more comprehensive interpretation of conventional TEM analyses and the results obtained from mechanical testing. To produce relevant and reliable complementary information to conventional imaging and destructive methods, numerous aspects need to be considered. This paper addresses the application of PAS techniques in the characterization of materials exposed to different types of irradiation experiments. It reviews recent studies in the field, and provides some empirical support for exploiting the full potential of positron annihilation spectroscopy in nuclear material research.

2. Methodology

Positron annihilation spectroscopy (PAS) refers to a qualitative and quantitative analysis of spectra acquired from different spectrometry techniques. The two most common techniques based on electron-positron annihilation are positron annihilation lifetime spectrometry (PALS) and Doppler broadening spectrometry of annihilation radiation (DBS). It is important to note that an overwhelming number of published studies, including the present study, refer to spectroscopy rather than spectrometry. The likely reason for this is that the aim of these studies is usually focused on the origin and properties of the acquired spectra rather than on the actual measurement. The unique feature of the positron as a probe for microstructural characterization is its remarkable sensitivity to crystal lattice defects, ranging in size from monovacancy to open-volume defects or precipitates of few nanometers in diameter. How efficiently the defect acts as a potential well for the positron depends mostly on the size of the defect (open-volume defects) and the positron affinity (precipitates—how different the positron affinity is to that of the matrix). The accuracy of determining the interaction volume of the positron depends on its energy. The selection of a suitable PAS technique depends on the actual phenomenon studied.

2.1. *Positron Stopping in Solids—Interaction Volume*

The availability of different radioisotope sources and variable-energy slow positron beams enables the positron annihilation-based experiments to be optimized to suit the material damage profile to be investigated. Although studies focusing on a particular type of positrons source and corresponding stopping profile have been published in the past [22–25], empirical studies employing both radioisotope positrons and slow positron beams in the same field of material research are rather scarce. In the material research studying radiation effects, the radioisotope source has been used typically in the context of neutron irradiation and gamma irradiation, i.e., deeply penetrating ionizing radiation. On the contrary, slow positron beams have been predominantly used in the studies employing charged particle accelerators, producing a typical narrow-peak damage profile in the irradiated samples. With the increasing energy of particle accelerators and availability of high-intensity ion beams of tens of MeV, the classical dividing between "near-surface" and "bulk" radiation damage studies loses its justification. In the following section, we summarize some fundamental knowledge on the interaction volume associated with positron stopping in solids.

2.1.1. Radioisotope Positron Sources

Beta-plus decay radioisotopes are the most common types of positron sources used in material research. In addition to the requirement of a reasonably long half-life (at least a few years), the emission of the positron must be accompanied by the emission of characteristic gamma to enable the application of techniques based on positron lifetime measurements.

It is important to note that regardless of the source design, a fraction of positrons will not escape from the source, and electron-positron annihilation signals will always contain a source contribution. The most common procedure for positron source preparation is based on the use of ^{22}NaCl solution evaporated and sealed in a Kapton encapsulation. Depending on the thickness of the Kapton foil and the effective diameter, the source contribution of such sources can vary from ~7 to ~20% [26,27].

Another type of positron source with various prospective applications is based on a ^{44}Ti/^{44}Sc radionuclide generator. Due to the beneficial mechanical properties of metallic foils, compared to brittle ionic bonds in ^{22}Na sources, no encapsulation is required to prevent sample contamination from the positron source. Since the production of ^{44}Ti requires a high-energy proton implantation [28] with limited availability of the information about actual reactions and reactions cross-sections, it is necessary to consider the contribution of positron annihilation inside the source as a function of its thickness. The actual source contribution is, however, also a function of the fraction of backscattered positrons. In the field of nuclear materials and testing of miniaturized samples, it is reasonable to outline the fraction of backscattered positrons as a function of the tested sample's thickness. Figure 1 shows the ^{44}Ti/^{44}Sc source contribution calculated using GEANT4 code [29] for three different source thicknesses and various thicknesses of the used sample (Fe). With a sufficiently thick sample, the (saturated) source contribution was estimated to be 3.2%, 21.2% and 56.2% for the titanium/scandium source in the form of a 5 μm, 25 μm and 250 μm thick foil, respectively. In the case of thinner samples, the contribution of the source is reduced by the fraction of positrons escaping the source-sample system (no backing materials considered in the simulations). Figure 1 also includes a commonly used positron source based on ^{22}Na encapsulated in 7.5 μm Kapton foil.

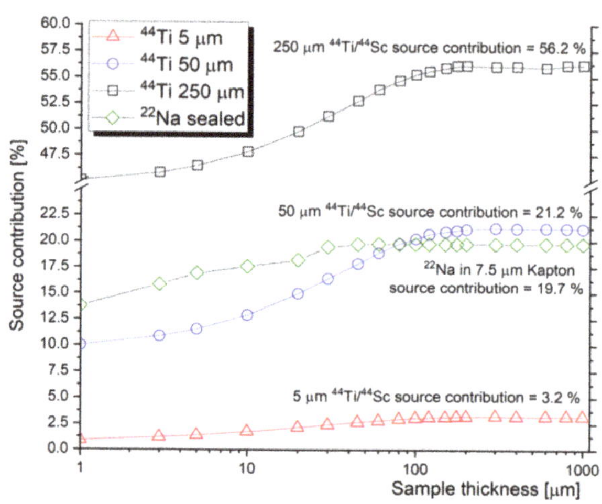

Figure 1. GEANT4 simulation of the source contribution for ^{44}Ti/^{44}Sc (different thickness of the source foil) and ^{22}Na (encapsulated in 7.5 μm Kapton foil).

Although the energy spectrum of positrons (i.e., the positron stopping profile) from any source depends on the encapsulation and the actual thickness of the source, positron mean stopping depth from the ^{44}Ti/^{44}Sc source is roughly 5× deeper than the stopping depth of ^{22}Na positrons. Figure 2 shows the estimated stopping profile of positrons from the two radioisotope sources in Fe. While the ^{22}Na source that is shown is encapsulated in Kapton, the ^{44}Ti/^{44}Sc is considered in the form of metallic foil with different thicknesses. The results are obtained from GEANT4 simulation for 7.5 μm Kapton-encapsulated ^{22}Na source and ^{44}Ti/^{44}Sc source. Note that the thickness of the source does not significantly moderate the energy spectrum, and the stopping profile of the positrons from different thicknesses of titanium foil is practically identical.

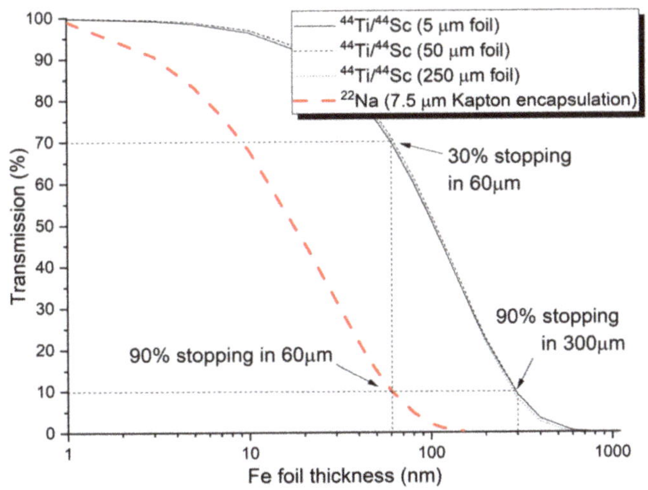

Figure 2. Stopping profile of positrons emitted from a ^{22}Na Kapton-encapsulated source (thick dashed line) and from ^{44}Ti/^{44}Sc foil sources of different thicknesses (thin lines).

The choice of optimal source depends on the nature of the positron characterization experiments as well as on the investigated material. While metallic sources are usually preferred for semiconductors studies, polymer-encapsulated sources provide a good choice for studies on metallic samples. The study by Saro et al. [30] concludes that practically measurable signal (20% positrons stopped in the modified region) from ion-implanted region can be obtained for ion beam modified layer of at least 3 μm thickness. This is a practical reference for mechanical surface treatment where the layer of plastically deformed material can reach up to 1 micrometer even when sub-micrometer diamond polishing slurry is used [31]. Therefore, electropolishing following careful mechanical polishing is strongly suggested as a surface preparation procedure for any experiments involving moderated or slow positron.

Several papers have been published on analytical characterization of positron stopping profile in different metals with a good correlation in the bulk region and certain discrepancies in the near-surface region [27,32]. These discrepancies may originate from low-energy positron backscattering (function of Z) and the actual design of the considered positron source.

2.1.2. Slow Positron Beam Experiments

A relatively wide interaction range of electron-positron annihilation in experiments utilizing radioisotope positron sources is not practically suitable for narrow ion beam modified layers. Various techniques based on slow positron beams have been developed to study thin multilayer systems or near-surface areas of ion-bombarded samples. While the positron yield of the sources in such facilities is below the yield of conventional radioisotope positron sources, there have been some high-intensity positron sources developed, such as the NEPOMUC source [33] at the FRMII reactor in Garching, Germany. A unique feature of such facilities is an absence of the "source component", providing an undisturbed spectrum characterizing the studied sample exclusively.

The techniques based on slow positron beams utilize the relationship between the energy of the moderated positron beam and the depth of the positron stopping. The mean value of the depth of the positron stopping profile, $\overline{z}$, is a function of the positron energy and target material density ρ (Equation (1)).

$$\overline{z}[\mathrm{cm}] = \frac{A E\mathrm{keV}^{n}}{\rho[\mathrm{g\ cm^{-3}}]} \quad (1)$$

The constants A = 40 g cm^{-2} keV^{-n} and n = 1.6 are empirically determined and applicable for a wide range of structural materials. The shape and width of the monoenergetic stopping profile can be described by the Makhovian formula (probability density function (2)), where z_0 is related to the mean implantation depth by $z_0 = \frac{2\bar{z}}{\sqrt{\pi}}$ and the shape parameter m = 2 [34].

$$P(z,\,E) = \frac{mz^{m-1}}{z_0^m} e^{-\left(\frac{z}{z_0}\right)^m} \tag{2}$$

While the Makhovian profile is relatively narrow for low positron energies (<10 keV), it broadens significantly for energies above ~15 keV, as can be observed in Figure 3 (note: log scale on y-axis). This a very important feature to be considered in the interpretation of slow positron beam data obtained on multilayer samples or ion-implanted samples; for instance, one sigma interval corresponding to Makhovian profile of 15 keV positron is 200 nm, with a mean implantation depth of 440 nm.

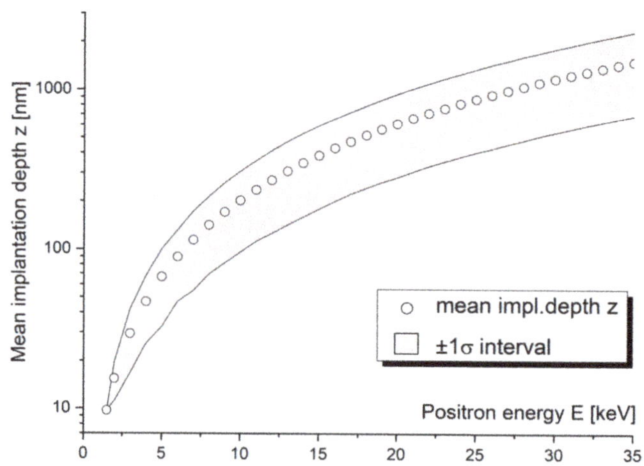

Figure 3. Mean implantation depth of positrons with variable energy (E). The standard deviation was calculated from the Gaussian-like Makhovian function as $\frac{FWHM}{2.3548}$, where FWHM is full width at half of the maximum [35].

2.2. *Techniques of Positron Annihilation Spectroscopy*

Various techniques of positron annihilation spectroscopy have been established in the past to provide unique information at the atomic scale on characterizing defects in crystalline materials both in qualitative and quantitative terms. The two most common techniques are positron annihilation lifetime spectroscopy (PALS) and Doppler broadening spectroscopy of the annihilation gamma peak (DBS) [36]. While the fundamental principles of the two techniques (described in detail below) are different, a good correlation can often be obtained by their complementary application [37]. Together with transmission electron microscopy (TEM) and some other microstructural characterization methods, such as SANS and Raman spectroscopy, PAS provides solid support for interpreting the results obtained from micromechanical testing methods.

2.2.1. Positron Annihilation Lifetime Spectroscopy (PALS)

Positron annihilation lifetime spectroscopy is a widely used analytical technique utilizing the phenomenon of positron trapping by defects and the lifetime of the positron being dependent on the nature and size of this defect. The values of bulk lifetimes and elementary defects are well known virtually for all basic metals and simple alloys, enabling a qualitative characterization of various defects introduced in irradiation, thermal or mechanical treatment. There are a few practical limitations of qualitative analyses by positron lifetime spectroscopy. First and foremost, the limited number of components into which the lifetime spectrum can be decomposed. There are rarely more than 2–3 discrete components determined, additional to the ones assigned to the source contribution. This problem is even more pronounced in experiments leading to combined defects (e.g., jogs or vacancies on dislocation lines) or a broad continuous spectrum of open-volume defect size (instead of a sharp unimodal or bimodal distribution) complicating the spectra decomposition. Although modern digital lifetime spectrometers [38] have excellent time resolution (FWHM < 150 ps), reliable decomposition requires a difference of at least several tens of ps in lifetimes of the identified components. Finally, it is essential to consider all contributions to the lifetime spectra outside the source sample sandwich, such as sample holder or positron backscattering to sample surface (or any internal surface), which act as a strong trapping site. The experiment geometry and good calibration samples are essential in this regard.

Figure 4 shows a typical 1M count lifetime spectrum of a ferritic/martensitic (f/m) steel sample in as-received condition and after irradiation in spallation neutron source to 20 dpa (2000 appm He) [37]. Besides, the figure shows the lifetime spectrum of pure defect-free well-annealed Fe, i.e., a two-component (Fe sample + source encapsulation) lifetime spectrum. One can observe a distinct increase in background as well as an increase in the contribution of longer lifetime components in the irradiated Fe–9Cr sample. This example shows a particular case when the trapping of radiation-induced defects leads to near-saturated positron trapping when the whole lifetime spectrum can be described by a single component (fitted by a single exponential curve) and the short lifetime component cannot be identified anymore. For more details on this, view the positron trapping model in Section 2.4.

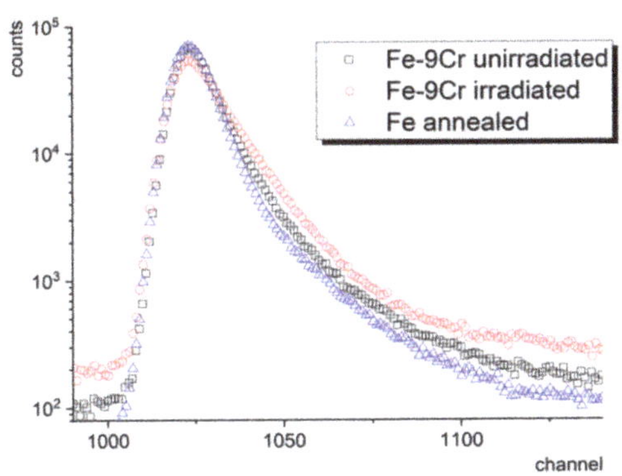

Figure 4. Lifetime spectrum of defect-free annealed Fe, together with the spectrum of Fe–9Cr steel before and after exposure to radiation environment of spallation neutron source [37].

2.2.2. Doppler Broadening Spectroscopy

Spectroscopy based on the measurement of the Doppler broadening of the annihilation peak is based on the momentum conservation law of the electron–positron pair and the fact that the contribution of the positron to this momentum is practically negligible. This enables us to observe the changes in the shape of the annihilation peak induced by changing ratio between low-momentum valence/conduction electrons and high-momentum core electrons. It also enables changes in the atom-specific core electron distribution to be detected and thus provides chemical information from the vicinity of the annihilation site.

In PALS spectra, the lower count region corresponds to a long positron lifetime, often not very meaningful in the research of metals and alloys, while the low count region of the Doppler-broadened annihilation line is very important for the investigation of the chemical fingerprint of the annihilation site. The wing parts of the spectra describe the contribution of core electron annihilations and, in a simple one-detector DBS spectrometer (gamma spectrometer), remain almost entirely masked by the background. The reduction in the background and increase in the peak-to-background ratio is therefore very important. A convenient, well-established way to obtain virtually no background DBS profile is to use two HPGe detectors facing opposite sides in a coincidence setup. This only enables the recording of events when two (annihilation) gammas are detected in a certain time window, i.e., time discrimination. The spectra of such a coincidence setup of DBS, referred to as CDBS, are recorded in the form of a two-dimensional matrix, which can be plotted as a three-dimensional image, as shown in Figure 5. The analysis of the spectra is performed using a diagonal indicated in the two-dimensional projection of the spectra.

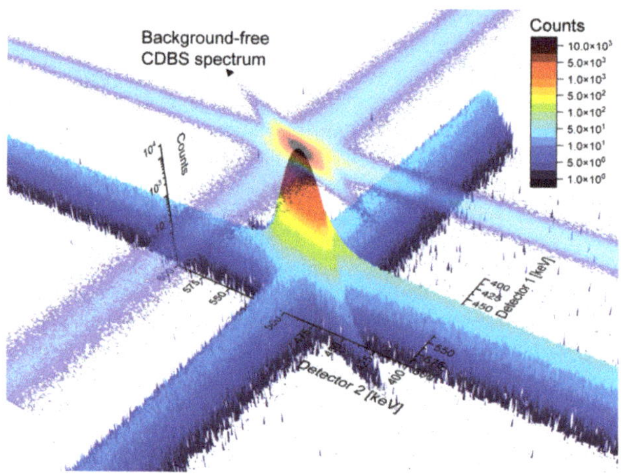

Figure 5. CDBS spectra obtained on defect-free Fe in a three-dimensional plot and two-dimensional projection. The elliptical region extending diagonally with $E_{D1} + E_{D2} = 1022$ keV is virtually background free.

Naturally, the count rate in 2D CDBS spectra is significantly (roughly an order of magnitude) lower compared to single-detector DBS spectra. The counts determine the actual time needed to acquire a spectrum in the high-momentum (low count) region. Our recent experiments on thermal-induced Cr precipitation in PM2000 steel, however, show that the best compromise between the sensitivity to the chemical changes in the microstructure and the count rate can be obtained from the 2D projection of the spectra to the X(Y)Z plane, i.e., when the coincidence spectrum is discriminated with respect to the time (the event detected by first detector is considered valid only when the second detector detects a signal within the given time window). In other words, the "massive" reduction in count rate in the diagonal region of the CDBS spectra is not always outweighed by the benefits of "massive" background reduction. Figure 6a shows normalized one-dimensional Doppler broadening spectra obtained as the main diagonal of the CDBS matrix (10^7 counts) and as the sum of the lines (columns), respectively. One can clearly observe an increased peak-to-background ratio in the main CDBS diagonal. On the other hand, the amount of counts in the summed lines (columns) is significantly reduced in the high-momentum region. This leads to a significant worsening of the uncertainty, calculated as 1/sqrt(counts), in this region. Figure 6b,c show DBS momentum curves relative to the pure Fe curve obtained for the two discussed cases. In spite of different uncertainty given by different statistics of counts, the shape of the ratio curves also varies between different data treatment procedures, as can be observed from the figures. Chemical interpretation of the CDBS data obtained on complex engineering materials, such as alloys and steels, must be performed with caution. It is important to note that more sophisticated CDBS spectra treatments are often used in similar studies, improving the uncertainty of the acquired data. It is, nevertheless, beneficial to consider the setup of the equipment, data treatment procedure and the target statistics of the spectra. Meaningful Doppler broadening data can be obtained from a coincidence setup utilizing one HPGe detector and one (much more affordable) scintillator detector. More detailed discussion on the processing of CDBS spectra, including reference to a new update of a CDBStools software developed at the Slovak University of Technology, can be found in a paper by Petriska et al. in this issue.

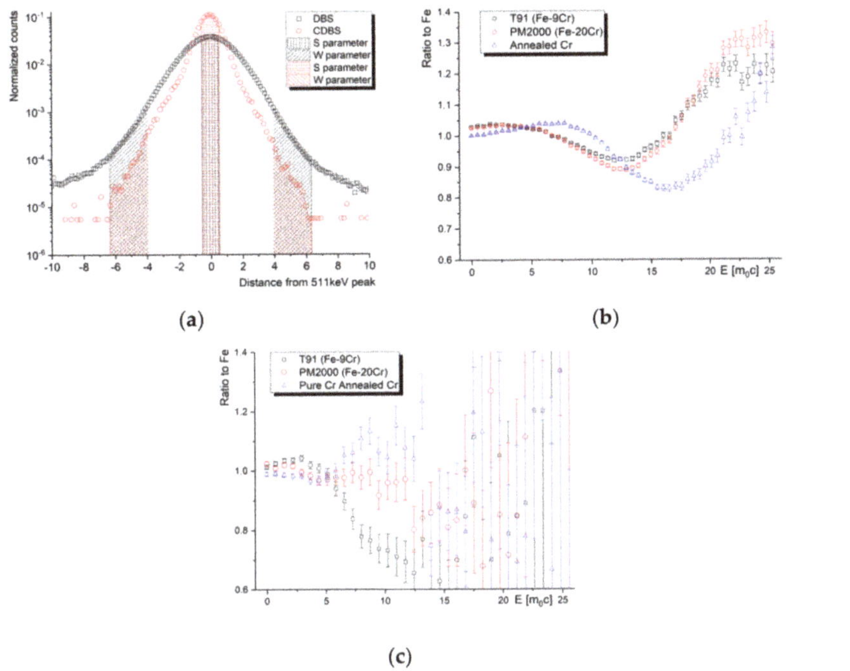

Figure 6. Normalized DBS spectra (**a**) of pure Fe obtained from CDBS matrix as the main diagonal (referred to as DBS) and as the sum of the lines (columns), respectively (referred to as CDBS). DBS ratio curves, relative to Fe, of two steels with different Cr content and pure Cr sample obtained for the first case (**b**) and for the second case (**c**).

As illustrated in Figure 6a, quantification of the broadening of the annihilation peak can be performed using two line-shape parameters, S and W, which are defined as the area under the central part of the peak and the area in the "wing" sections in a background-subtracted normalized DBS curve. This approach is widely used in most of the studies aimed at the "quantification" of radiation damage in nuclear materials or efficiency of recovery annealing in irradiated materials. In the case of strong positron sources and a sufficient number of counts in the high-momentum (wing) parts of the peak, more precise insights into microstructural changes can be obtained by evaluation of the actual momentum curves plotted as a ratio to the reference sample, as illustrated in Figure 6b,c. In the case of nuclear structural materials, the reference sample is usually a well-annealed defect-free Fe or a sample of the given studied material in the as-received condition.

2.3. *Positron Trapping at Defects*

In a perfect lattice, all positrons annihilate in the free delocalized state. The introduction of defects into the perfect lattice leads to localization of the positrons at these defects. In positron lifetime spectroscopy, it is represented by an introduction of a new (longer) defect component (τ_d) and a reduction in the free positron lifetime, typically represented by the shortest component of the spectra (τ_1). In the case of one type of defect, this can be explained using a simple trapping model, as follows:

$$\frac{1}{\tau_1} = \frac{1}{\tau_B} + \kappa_d \quad (3)$$

where the following applies:

$$\kappa_d\left[s^{-1}\right] = \frac{I_2}{I_1}(\lambda_b - \lambda_d) = \frac{I_2}{I_1}\left(\frac{1}{\tau_b} - \frac{1}{\tau_d}\right) \quad (4)$$

is the positron trapping rate at defect d, and $\lambda_b = \frac{1}{\tau_b}$ and $\lambda_d = \frac{1}{\tau_d}$ are the annihilation rates in bulk and in defects, respectively. Intensities I_1 and I_2 represent the fraction of positrons corresponding to either group. Considering a bimodal characteristic and two types of radiation-induced defects, which is a reasonable approximation, for instance, to the critical size concept of bias-driven cavity growth [39], the lifetime spectra can be analyzed in terms of three exponential components [4], as follows:

$$n(t) = n_0 \sum_{i=1}^{3} I_i exp\left(-\frac{t}{\tau_i}\right) \quad (5)$$

where n is the number of positrons at time t, n_0 is the number of positrons at time $t = 0$, and τ_i and I_i are the lifetimes and intensities of individual components, respectively. While τ_1 and I_1 describe the annihilation with delocalized (untrapped) positrons, τ_2, I_2, τ_3, I_3 refer to positrons trapped at defects.

In numerous works, including many of those reviewed in this paper, the (total) positron trapping rate at defects is given as a sum of the trapping rate in the defect of type 1 (κ_1) and the trapping rate in the defect of type 2 (κ_2).

$$\tau_2 = \frac{1}{\lambda_2} = \frac{1}{\lambda_1};\ I_2 = \frac{\kappa_1}{(\lambda_B - \lambda_1 + \kappa_1 + \kappa_2)} \quad (6)$$

$$\tau_3 = \frac{1}{\lambda_3} = \frac{1}{\lambda_2};\ I_3 = \frac{\kappa_2}{(\lambda_B - \lambda_2 + \kappa_1 + \kappa_2)} \quad (7)$$

The positron trapping rate at defects, κ [s^{-1}], is directly proportional to the density of the given type of defects via the so-called positron trapping coefficient μ [m^3s^{-1}]. This constant has been obtained for point defect and small defect clusters in a wide range of materials. It is important to note that the trapping rate is proportional to the density of traps only in so-called transition-limited positron trapping, which assumes a small size of positron traps with a spatial extent that is much less than the positron thermal wavelength (~few nm). This assumes homogeneous distribution of (small) defects, zero probability of detraining from defects, as well as zero probability of trapping of non-thermalized positrons. In the case of large defect agglomerations (>2–3 nm), such as bubbles or voids, positron diffusion needs to be considered, and consideration of diffusion-limited trapping [40] is advised. In constant-temperature measurements (constant thermal velocity of positrons), the transition from the transition-limited to diffusion-limited regime in the irradiated samples usually arises from the lower cavity concentration and the larger trapping coefficient due to the larger cavity sizes.

In the case of more than one type of defect present in the investigated sample, it is reasonable to consider the evaluation of the average positron lifetime (τ_{AVG}), rather than the lifetimes and intensities of the individual components of the spectra. τ_{AVG} is defined as a weighted average of positron lifetime components. According to the simple trapping model, with increasing concentration of defects, the positron trapping at defects is enhanced, which results in an increase in τ_{AVG}.

The introduction of defects is, in terms of DBS measurements, typically accompanied by an increase in the line-shape S parameter, which characterizes the Doppler broadening of the annihilation line due to the non-zero momentum of the annihilating electrons. Unlike the positron lifetime carrying information about the local electron density at the annihilation site, the S parameter carries information about the local electron momentum distribution. This feature is particularly valuable in the case of production of transmutation inert gasses, such as helium or argon, in harsh radiation environments or in the case of radiation-induced precipitation processes.

2.4. Positron Diffusion in Solids

For the efficient treatment of the data obtained from the slow positron beam measurements, not only the initial implantation profile, but also the thermalized positrons diffusion should be considered. This is considered in the positron diffusion trapping model (DTM) [23].

In the case of so-called transition-limited trapping assuming homogeneously distributed (small) defects, the positron diffusion constant D_+ is not a function of space coordinates, and the effective positron lifetime τ_{eff} is proportional to the effective positron diffusion length in the given material (Equation (8)).

$$L_{+,eff} = \sqrt{D_+ \tau_{eff}} \tag{8}$$

$L_{+,eff}$ can be obtained from slow positron experiments utilizing monoenergetic beams of positrons with variable energy. For this purpose, the VEPFIT software [41], fitting of the curves, is traditionally used. This program numerically solves the positron diffusion equation, and provides positron diffusion lengths and ranges of individual layers in multi-layer systems. However, our recent experiment [42] on slow positron lifetime measurements suggests that the same approach can be used for $\tau_{AVG}(E)$ curves obtained from average positron lifetime data. This study builds on numerous experiments showing an excellent correlation between the (E) and $\tau_{AVG}(E)$ obtained on helium-implanted steel samples.

Using Equations (3) and (8), one can express the positron trapping rate at defects [s^{-1}] as a function of positron diffusion length by the following formula:

$$\kappa = \frac{1}{\tau_B}\left(\frac{{L_{+,B}}^2}{{L_{+,\,eff}}^2} - 1\right) \tag{9}$$

Here, $L_{+,B}$ refers to the positron diffusion length in defect-free bulk. The positron trapping rate can be obtained by Equation (9), even in the case of saturated positron trapping when the positron lifetime does not reflect the increased concentration of defects anymore. This approach is very beneficial in studies of irradiated materials with a high density of internal surfaces and other defects, such as tempered martensitic steels and nanostructured alloys.

3. Application of PAS Techniques in Accelerator-Based Irradiation Experiments

The following section reviews some of the recent papers published on the PAS characterization of ion-implanted structural materials used or considered in nuclear applications. More than 30 papers, selected from scientific libraries by using the keywords "positron annihilation" and "ion implantation", have been reviewed and complemented by our studies and experiments. The presented discussion was divided into four sections, based on the type of irradiation experiment, namely, self-ion irradiation, hydrogen (proton) irradiation, helium (alpha) irradiation, and combined synergistic studies utilizing at least two of these types of ions.

3.1. Self-Ion Irradiation

The primary benefit of using heavy-ion or self-ion irradiation lies in the high displacement damage rate and the ability to obtain high dpa's without a significant change in the chemical composition of the irradiated material. It is important to note that a significant fraction of the implanted damage is "lost" in the heat spikes, and the final amount of damage produced may be much less than the estimation provided by, for instance, the SRIM code [43], as this is limited to the binary collision approximation, which does not describe defect recombination. A better prediction is provided by, e.g., models based on the two-temperature model [44], with extensions focusing on the interaction with metals [45], or in combination with molecular dynamics, e.g., [46]. As in swift heavy ions, the high electronic stopping (strong inelastic electronic excitation) characterizing the initial part of the ion trajectory significantly dominates over the nuclear stopping (elastic nuclear collision). This type of irradiation experiment usually results in a relatively narrow region of ion beam modified material. From the positron annihilation perspective, the use of unmoderated radioisotope sources is practically not applicable here, and the whole damage region can be characterized by slow positron beam experiments with a positron energy of a few keV. On the one hand, the corresponding Makhovian profile corresponds to a very localized positron–sample interaction volume, while the quality of the sample surface, on the other hand, affects the obtained results significantly. Unless very careful sample preparation, finished by electrochemical polishing, is performed, the near-surface region should be described by a separate layer in the positron diffusion model. The interpretation of the positron annihilation spectroscopy data obtained on self-ion implanted materials is usually much more straightforward compared to the experiments discussed in the next chapter. Without transmutation elements, such as H or He, the positron lifetime increases monotonically with the size of the (vacancy-type) defect cluster. This enables a rather reliable qualitative characterization of the radiation-induced defects. There is also a typical linear dependency between the DBS line-shape parameters S and W, indicating one type of defect with no changes in the chemical environment at the annihilation site. The typical energy used in the self-ion irradiation experiments characterized by slow positron beam techniques, published in the literature, lies in the range of a few MeV, i.e., in the sub-μm region (Table 1).

Table 1. Selected PAS studies published on self-ion irradiation of nuclear structural materials.

Material	Ion (E)	T [°C]	Fluence(s) [cm^{-2}]	Techniques	Ref.
FeCrCoNi (HEA)	Ni^+ (1.5 MeV)	500	5×10^{16}	DBS (SPB 0.25–21.5 keV) + VEPFIT analysis; GIXRD; EBSD	[47]
Fe film	Fe^+ (2 MeV)	RT	5.65×10^{14}	DBS (SPB 0.05–16 keV) + VEPFIT analysis; LT (SPB 0.5–16 keV)	[48]
BCC Fe	Y^+ (1.2 MeV)	RT	1.0×10^{14} 2.0×10^{15} 3.0×10^{15}	DBS (SPB 0.5–25 keV) + VEPFIT analysis, SIMS	[49]
Fe–9%Cr (T91 steel)	Fe^+ (3.25 MeV)	RT; 300; 450	1.7×10^{16}	DBS (SPB 0.5–20 keV) + VEPFIT analysis; nanoindentation	[50]
High Si Fe–11Cr	Fe^+ (3.25 MeV)	RT, 450	4.3×10^{15}; 1.7×10^{16}	DBS (SPB 0.18–20.18 keV) + VEPFIT analysis	[51]
RPV steel (JRQ)	Fe^{2+} (5 MeV)	300	2.66×10^{13} - 2.66×10^{15}	DBS (SPB 30 eV–36 keV); Nanoindentation	[52]

In the work by Abhaya et al. [47], Ni^+ ions of 1.5 MeV and fluence of 5×10^{16} cm^{-2} were used to produce damage of nearly 100 dpa in an FeCrCoNi high-entropy alloy at 500 °C. The irradiation led to the creation of monovacancies, which were observed to be stable up to 700 °C. The authors concluded that there was a significant effect of the grain boundaries in the polycrystalline FeCrCoNi, acting as strong sinks and preventing the formation of voids, which would be expected at the given displacement damage level and at the used irradiation temperature.

The slow positron beam experiment, utilizing both positron lifetime and Doppler broadening techniques, has been used in Agarwal et al. [48], in the characterization of a mechanism for the interaction of cascade damage with voids in Fe films. Ion irradiation of the Fe films was performed with 2 MeV Fe ions at room temperature, to a fluence of 5.65×10^{14} cm^{-2}. The relatively low irradiation fluence (0.06 dpa in peak) resulted in the formation of a maximum of four vacancy clusters assigned to a 258 ps lifetime component. The study concludes an interesting result, where the density of small vacancy clusters increases with irradiation, whereby the size of large voids is reduced, which is an effect attributed to the high porosity of the initial microstructure. However, the authors also note the limited accuracy of defect density calculations, due to missing positron trapping coefficients for small vacancy clusters.

Another PAS study devoted to Fe, by C.W. He et al. [49], used SPB DBS to investigate the interaction between vacancies and Yttrium atoms in BCC iron irradiated with 1.2 MeV Y ions at room temperature and fluences from 1×10^{14} cm^{-2} to 3.0×10^{15} cm^{-2}. The authors also observed a decrease in the size of the vacancy clusters with increasing fluence; however, in this case, the concentration of these decreased too. This effect is explained as a combination of the (i) formation of V_m–Y_n complexes, as the migration temperature of the vacancies in BCC Fe is below RT and the mobility of the vacancies can be further reduced by binding with Y, which limits their agglomeration; (ii) formation of Y_m–X_n precipitates (where X = Y,O, etc.), leading to a decrease in the size and concentration of vacancy clusters.

Commercial-grade 9Cr–1Mo steel (T91) was investigated by SPB DBS after implantation by 3.25 MeV Fe ions at different temperatures, by Zhu et al. [50]. The applied DBS technique revealed that the concentration of open-volume defects decreased with the implantation temperature, while the complementary nanoindentation technique showed an increase in hardness with increasing irradiation temperature. Similar irradiation conditions were used in another work of Zhu et al. [51] on the NHS, a high Si reduced activation f/m steel. In both the fluences used in this research, the S parameter was found to increase significantly after RT irradiation. On the other hand, the irradiation performed at 450 °C resulted in an S-parameter depth profile that was almost identical to the one of the unirradiated sample. This was explained by the enhanced mobility of the vacancies and interstitials at high temperatures, together with the concentration gradient of SIA, due to excess Fe atoms from the deeper layers (while having a vacancy-rich ion-track region).

The work of Pecko et al. [52], on RPV steel JRQ containing 0.15 wt.% of Cu, addressed some empirical connections between Cu atoms and the vacancies introduced by self-ion Fe irradiation (Fe^{2+} 5 MeV; at 300 °C; 2.66×10^{13}–2.66×10^{15}). While the vacancy-type defects were found to not be directly responsible for the radiation-induced hardening of the steel, they take part in the radiation-enhanced diffusion of Cu atoms and the formation of Cu-rich clusters.

3.2. *Hydrogen (Proton) Implantation*

In many ways, proton irradiation overcomes the drawbacks of using heavy-ion irradiation as a surrogate for neutron irradiation. Accelerators that are capable of delivering protons of a few MeV are relatively widely available, enabling the modification of tens of micrometer-thick layers in the studied samples by relatively flat damage profiles [1]. The principal benefit of using proton irradiation is the small mass of the proton compared to heavy ions, resulting in a lower energy of recoils. This provides smaller, more widely spaced cascades compared to heavy ions or fast neutrons. Since a relatively small energy of protons is required to overcome the Coulomb barrier, there is often a chance to introduce the activation of certain samples, and this chance increases with proton energy [1].

After losing energy to the electronic degrees of freedom of the metal, a fast proton slows down to thermal velocities and captures an electron to become a hydrogen atom [53]. It is well known that the accumulation of hydrogen can lead to severe embrittlement after reaching a critical concentration near a crack nucleus [1]. Due to the small mass of the proton and the relatively low number of displaced atoms after collision with the target material, the irradiation fluences must be relatively high in order to introduce distinct radiation damage. This opens up the question of the solubility and mobility of hydrogen in the target materials, and the possible effect of blistering. Due to the extremely low activation energy of hydrogen diffusion in alpha iron (0.045 eV), the threshold for blistering in Fe is very high (8×10^{18}–1.5×10^{20} cm^{-2}) [54], and ferritic steels can accommodate very high fluences of implanted hydrogen. Another consequence of the high mobility of hydrogen in Fe-based materials is that the Bragg peak is smeared out in-depth by diffusion [54], i.e., the SRIM profiles cannot be simply correlated by the results of, for instance, VEPFIT analysis.

The effect of hydrogen cannot be neglected in any proton irradiation experiments. The diffusion of hydrogen into radiation-induced open-volume defects does not only change the susceptibility of the material to brittle fracture, but it also strongly affects the process of positron trapping at these defects. The estimation of the hydrogen content in the open-volume defects is further complicated by the presence of fast migration paths, such as grain boundaries, lath boundaries, dislocations, etc., which are typical for ferrous alloys and steels.

In the technique of the measurement of Doppler broadening of the annihilation line, the results (line-shape parameters S and W) are often presented in the form of a so-called S-W plot. The linearity of the data in such a plot usually indicates one type of defect in the matrix, with no change in the chemical composition at the vicinity of the annihilation site. Many authors report a change in the linearity (slope) of the S-W data after proton implantation, attributing this trend change to the effect of hydrogen. There are, however, numerous proton irradiation experiments reported with no indication of a change in the S-W trend line. A well-known effect of hydrogen on the lifetime of positrons trapped in vacancy clusters was reported for various materials by theoretical modelling [55–57]. The experimental validation of the lifetime data is, however, very complicated for irradiation experiments, leading to the formation of vacancy clusters, which practically involves all types of irradiations except electron and low-energy proton irradiation (which do not lead to the creation of defects larger than monovacancies). Despite these hurdles, proton irradiation combined with post-irradiation examination (PIE), using positron annihilation techniques, provides extremely valuable complementary data for the research of the radiation tolerance of nuclear structural materials. Some selected papers utilizing this approach are summarized in Table 2 and discussed below.

Table 2. Selected PAS studies published on proton irradiation of nuclear structural materials.

Material	Ion (E)	T [°C]	Fluence(s) [cm^{-2}]	Techniques	Ref.
Fe–9Cr alloy	H^+ (11 MeV); H^+ (150 MeV), incl. H (1100 appm/dpa) and He (120 appm/dpa)	RT (150 MeV); 300 (11 MeV)	4.5×10^{15} (150 MeV); 6.0×10^{16} (11 MeV);	PALS (^{22}Na); Tensile	[15]
Polycrystalline Fe	H^+ (73 keV; 173 keV; multi-energy 50–173 keV)	RT	3×10^{15}; 1×10^{16}; 3×10^{16};	DBS (SPB 0.1–36 keV) + VEPFIT analysis; Elastic Recoil Detection (ERD)	[58]
Modified 310S steel	H^+ (50 keV);	290	4.0×10^{16}; 1.2×10^{17};	DBS (SPB ~ 0.2–20 keV); TEM	[59]
German RPV weld metals	H^+ (100 keV)	<100	6.2×10^{17}; 5.1×10^{18}; 2.0×10^{19};	PALS (^{22}Na)	[60]
Chinese RPV steel (A508-3)	H^+ (240 keV)	<100	2.5×10^{16}; 5.5×10^{16}; 1.1×10^{17}; 2.5×10^{17};	DBS (SPB 0.25–26 keV); Nanoindentation	[61]
Fe–9%Cr steel (T92)	H^+ (250 keV)	RT	0.01 dpa; 0.05 dpa; 0.2 dpa	DBS (SPB 1–25 keV); TEM; Nanoindentation	[62]
SA-738Gr.B steel (AP1000 reactor containment steel)	H^+ (400 keV)	150	1.07×10^{17}; 2.68×10^{17}; 5.37×10^{17};	PALS (^{22}Na);TEM	[63]

An excellent comprehensive study on a high-energy proton-irradiated Fe–9Cr alloy was published by Xu et al. [15]. Compared to the above-referenced studies, the work by Xu et al. involves proton energies above the spallation reaction threshold, inducing additional transmutation-based hydrogen and helium production, and spallation neutron damage. The positron lifetime of 137.7 ps that was measured for the unirradiated sample was attributed to dislocations and complex defects (vacancies trapped by dislocations), as well as to a carbide component. After irradiation, the long lifetime actually decreased, which was interpreted by the trapping of helium and hydrogen atoms on the above-mentioned defects. A similar reduction in the yield stress and tensile strength, i.e., softening of the alloy, was also explained by interstitial helium (150 MeV irradiation) and hydrogen atoms (11 and 150 MeV irradiations), as well as by the annihilation of jogs on dislocations (11 MeV). Only a few studies involving a proton implantation-induced nuclear reaction (spallation neutron production), and substantial production of transmutation hydrogen and helium in the irradiated samples, were published in the past, employing positron annihilation techniques [37,64,65]. However, as the overview of spallation neutron sources is not the subject of this review, this area of irradiation experiments is not discussed herein in more detail.

Horodek and Kulik used the variable-energy slow positron beam technique of Doppler broadening spectroscopy to characterize proton-implanted pure polycrystalline Fe [58]. Their results provide a good reference for more complex materials, such as alloys and steels modified by proton beams. Using the VEPFIT software, the authors obtained a positron diffusion length, in the reference material, of 158 ± 3 nm. This diffusion length is drastically reduced by proton implantation, to a few nm, regardless of the energy and fluence used in the experiment (see Table 2). The performed VEPFIT analysis further showed that the quasi-homogeneous profile obtained by multi-energy proton implantation can be effectively described by a two-layer model characterizing the implanted region and the unimplanted bulk (substrate) layer. More surprisingly, the two-layer VEPFIT model described single-peak implantation very well using an energy of 73 keV or 173 keV, respectively [58].

Proton-implanted samples of modified 310S steels were characterized by means of slow positron Doppler broadening spectroscopy and transmission electron microscopy in the study by Zhang et al. [59]. Irradiation at 290 °C, using 50 keV H^+ ions to fluences of 4.0×10^{16} and 1.2×10^{17} cm^{-2} led to a more pronounced production of hydrogen–vacancy clusters (referred to as proton–vacancy clusters) in Nb-, Ta-, and W-added steels, compared to Zr-added steels. Supported by TEM results on the size of dislocation loops, the authors concluded that the addition of oversized solute Zr to 310S steel leads to a more effective radiation resistance improvement than the addition of Nb, Ta, and W.

An interesting comparison of the proton-implanted samples of German RPV weld materials with their neutron-irradiated counterparts was published by Pecko et al. [60]. Identical levels of proton and neutron fluences were proposed in this experiment, but the relatively low energy of proton implantation resulted in a loss of some information from the conventional ^{22}Na positron lifetime measurements, with a positron stopping profile that was well below the introduced damage peak. Nevertheless, the increase in the positron lifetime was found to be roughly the same in the two irradiation experiments (slightly more pronounced for the proton-implanted samples). According to Pecko et al., neither of the irradiation experiments resulted in the formation of large vacancy clusters, with possible consequences in the form of embrittlement of the materials. Again, it is important to note that the use of the conventional radioisotope positron source for the 100 keV proton-implanted samples could not provide a sufficient signal from the implanted layer, and the data were, to a large extent, blurred by the positron annihilation in the unirradiated bulk.

In order to understand the nature of the hardening after radiation in reactor vessel steels, Chinese A508-3 steels were implanted by protons, and investigated by SPB DBS and nanoindentation techniques [61]. A good correlation observed between the two experimental techniques indicates a certain level of retention of hydrogen in the implanted samples, in the form of hydrogen–vacancy complexes, affecting the mechanical properties of the irradiated steel. On the other hand, the obtained S-W plot does not indicate notable changes in the data after proton irradiation, suggesting that there are changes in the nature of the defects. It is important to note that the implantation energy of the H^+ ions in [61] was 240 keV, in contrast to the 50–173 keV reported in [58], which resulted in most of the positron annihilation data obtained from the region being well below the ion accumulation peak.

The microstructural evolution induced by thermal ageing and proton irradiation was studied by transmission electron microscopy and positron annihilation spectroscopy in the work of Zhao et al. [62]. The study shows that proton irradiation leads to higher densities and larger sizes of dislocation-type defects, but less vacancy-type defects in the T92 steel samples aged at 650 °C (15,000 h) compared to the normalized and tempered reference steel samples. An interesting finding of the experiments was the linearity of the S-W plot, which was more affected by thermal ageing than by proton irradiation. This suggests the creation of some new phases, such as precipitates changing the chemical environment in the vicinity of the annihilation site.

Proton irradiation of structural materials, using energies of 400 keV and more, opens up the potential of applying conventional radioisotope sources as an alternative to slow positron beams. In the study by Ma et al. [63], relatively high fluences of H^+ ions (1.07×10^{17}–5.37×10^{17} cm^{-2}) were used to produce radiation defects in AP1000 reactor containment steel. Although the proton beam modified layer was clearly less than 3 μm thick, an increase in the positron lifetime was observed and used in the complementary characterization, along with TEM. The reason for the observed contrast in the positron lifetime spectra is likely the presence of large voids of several nm, identified by the TEM technique. These can act as strong sinks for positrons, providing a long lifetime component in the lifetime spectra (although the corresponding intensity could be relatively low).

3.3. Helium (Alpha) Irradiation

Helium has a detrimental effect on the evolution of radiation damage in steels, with a severe consequence for fission, fusion, and, particularly, accelerator-driven systems (ADS). Many radiation damage studies have been published on nuclear materials, involving helium (alpha) irradiation and various irradiation, and PIE experiments.

The argument against single-beam helium implantation is often the ratio of helium number density to displacement damage (dpa). The helium production rates in some fission applications are well below 1 appm/dpa [66]. In fusion applications, the ratio is on the level of a few tens of appm He/dpa [67], while, in spallation neutron sources, the helium production rates can reach 100 appm/dpa [68]. On the other hand, direct helium implantation (hundreds of keV ion energy) leads to a damage profile with a minimum level of a few hundreds of He appm/dpa, extending to a maximum of a few at.%/dpa. In our earlier study [69], we concluded, however, that the effect of the helium production rate (c_{He}/dpa) is not very significant at a helium concentration below a few thousand He appm. A comparison of helium-implanted f/m steels and spallation neutron source-irradiated f/m steels indicated the similar size of cavities (likely helium–vacancy clusters) at similar helium concentrations in the irradiated material.

The slow positron beam data obtained from the "flat" part of the helium implantation profile are widely considered as relevant for fusion and spallation applications, while, moving on to the helium concentration peak, the information obtained from the depth profiled measurements becomes devalued by an extremely high helium concentration and broadening of the positron stopping profile. Let us, nevertheless, look at an example of high-fluence helium implantation of the Fe–9Cr alloy, studied by the slow positron lifetime technique. Two similar materials were irradiated by helium, either in single-energy (250 keV) or subsequential two-energy (250 + 100 keV) implantation, to doses ranging from 1×10^{18} to 6.2×10^{18} cm^{-2} (Figure 7).

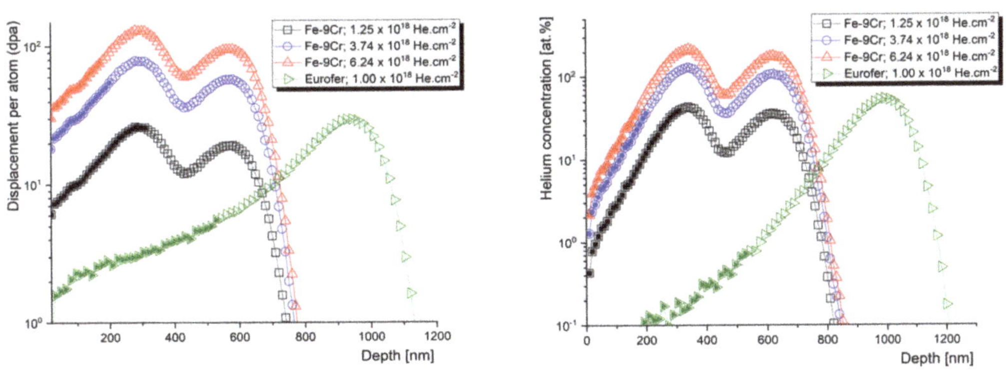

(a) (b)

Figure 7. SRIM calculated displacement damage (**a**) and helium concentration (**b**) profiles proposed for slow positron lifetime experiments.

Despite their low added value, most of the authors almost exclusively perform the evaluation of the slow positron lifetime data by plotting them as a function of the depth of the sample. Let us compare the information given by the positron lifetime depth profiles with the same data evaluated with respect to the irradiation conditions. The average positron lifetime, as the, statistically, most reliable parameter, is plotted as a function of the mean positron stopping depth in Figure 8. These data can be correlated to the SRIM profiles in Figure 7. Despite very different damage and c_{He} profiles, two Fe–9Cr alloys, with very distinct microstructures (Fe–9Cr binary and ODS Eurofer97), show almost identical depth profiles of the average positron lifetime (τ_{avg}) change (max. increase of

~30%) when irradiated to fluences of 1.0×10^{18} cm^{-2} and 1.25×10^{18} cm^{-2}, respectively. On the contrary, the positron lifetime increases much more significantly in the non-ODS Fe–9Cr alloy (Eurofer97) implanted by He to 1.0×10^{18} cm^{-2} (max. increase of ~55%). With increasing the total He fluence to 3.7×10^{18} cm^{-2} and 6.2×10^{18} cm^{-2}, the average positron lifetime in the binary Fe–9Cr alloy increases by 60% and 90%, respectively. In the first approach, we must consider three factors affecting these profiles. The first is the presence of defects sinks (such as grain boundaries, lath boundaries, dislocations, internal surfaces, etc.) in the pristine material. The second is the formation of new defect sinks (such as cavities) by agglomerations of radiation-induced defects. These cavities become primary sinks for new point defects and helium ions at high concentrations, and significantly suppress further growth of the radiation-induced cavities. Finally, the increase in positron lifetime in the given irradiated materials naturally depends on the fluence (number density) of the implanted helium ions.

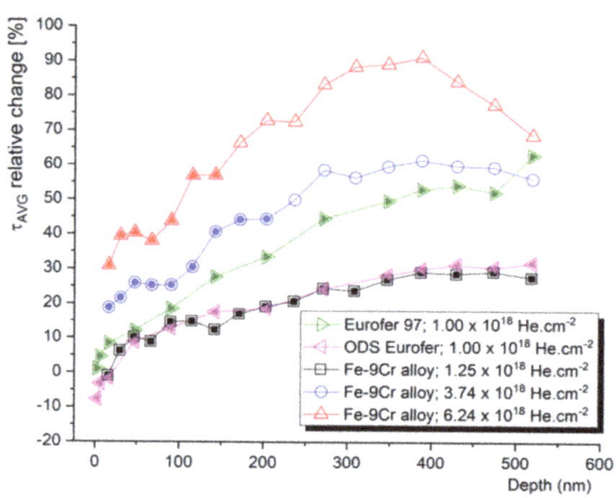

Figure 8. The increase in positron lifetime (implanted–unimplanted) as a function of the mean positron stopping depth. Filled symbols represent data with <50 at.% He, considered in further evaluations.

The depth profile of the average positron lifetime, such as the one in Figure 8, often causes scientists to incline to correlate it with SRIM profiles. This is, however, quite meaningless above a certain level of helium fluence/concentration, due to the accumulation of helium, blistering, and following exfoliation of the ion beam modified layer. Given the two energies of helium implantation used and the performed scanning electron microscopy (SEM) analysis, we can conclude that the blistering threshold for helium in the Fe–9Cr alloy is somewhere between 1.3 and 3.5×10^{18} cm^{-2}. At a total He fluence of 3.74×10^{18} cm^{-2}, about 15% of the surface was exfoliated, while at the fluence of 6.24×10^{18} cm^{-2}, about 25% of the surface was exfoliated after the implantation. The relative change (increase) in positron lifetime was multiplied by 1.15 and 1.25, respectively. The reason for this was to provide a more accurate estimation of positron lifetimes in extremely damaged Fe alloys. Nevertheless, in further evaluation, the peak concentration data were not considered, and irradiation conditions with a maximum of 40–50 at.% of helium were plotted in further graphs. These ranges of damage profiles are indicated by full symbols in Figures 7 and 8.

Figure 9 shows the relative values (implanted/un-implanted) of the average positron lifetimes obtained for the helium-implanted samples, and plotted as a function of c_{He} and c_{He}/dpa, respectively. In contrast to slow positron lifetime data evaluated with respect to the (mean) stopping depth of the positron, one can observe specific irradiation conditions leading to the given positron lifetime change. Although the number density of helium (c_{He}) seems to be the most representative irradiation parameter, it is important to not detach the c_{He} from the displacement damage (dpa). Since different energy of incident ions leads to a different number of primary knocked-on atoms (PKA), the information of displacement damage must be considered in the data interpretation. Figure 9 suggests that the increase in positron trapping at the defects increases rapidly at lower helium concentrations (lower dpa), and this increase slows down near saturation at high helium concentrations. It is reasonable to assume that this saturation is caused by new radiation-induced sinks for point defects and helium. At low c_{He}, the primary sinks for radiation-induced defects are the initial lattice defects (such as grain boundaries, lath boundaries, dislocations, internal surfaces, etc.), while at certain He fluences, the sinking of point defects takes place primarily at helium bubbles and other types of radiation-induced cavities. This must be considered in the experiments leading to the production of a high density of such cavities, i.e., effective defect sinks, which can give a certain impression of good radiation resistance.

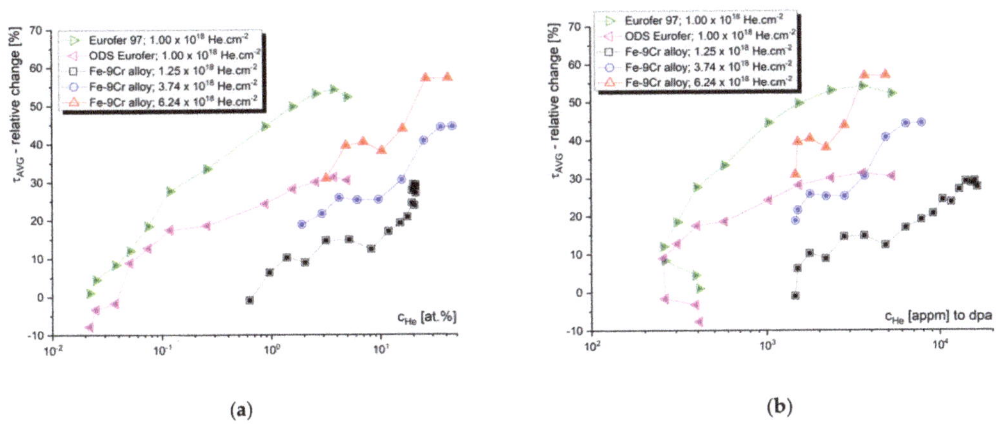

Figure 9. Relative increase in the average positron lifetime as obtained for the investigated samples. Data are plotted as a function of helium concentration c_{He} (**a**) and as a function of helium/dpa ratio (**b**).

Although τ_{avg} is the most reliable parameter statistically and its variation can be used for a rough assessment of the radiation tolerance of a given material, the comparison of absolute positron lifetime values between different materials and/or different irradiation conditions is far from being straightforward. It is, however, the aim of this paper to provide some empirical suggestions for researchers, regarding the interpretation of experimental PAS data in the research of the radiation resistance of materials. Let us plot the relative τ_{avg} data from Figure 9 as a function of the corresponding average lifetime obtained for unirradiated material (τ_{avg} un-implanted). As shown in Figure 10, such a relationship is fairly similar for the three Fe–9Cr alloys studied, which were irradiated to a similar fluence (1–1.25 $\times$ 10^{18} cm^{-2}). As expected, the most pronounced change (increase) in the positron lifetime was obtained for measurements where the τ_{avg} of the un-implanted sample was the shortest. In other words, the presence of trapping sites for positrons in the unirradiated material can be correlated with the presence of sinks for radiation-induced defects, i.e., the potential for local defect recombination. It is not surprising that the high positron lifetimes of the SPB data correspond to the surface and near-surface region (Figure 10b). This underlines the importance of considering the "surface effect" (positron diffusion to the surface), which acts both as a strong positron trap, and a sink for radiation-induced defects and helium.

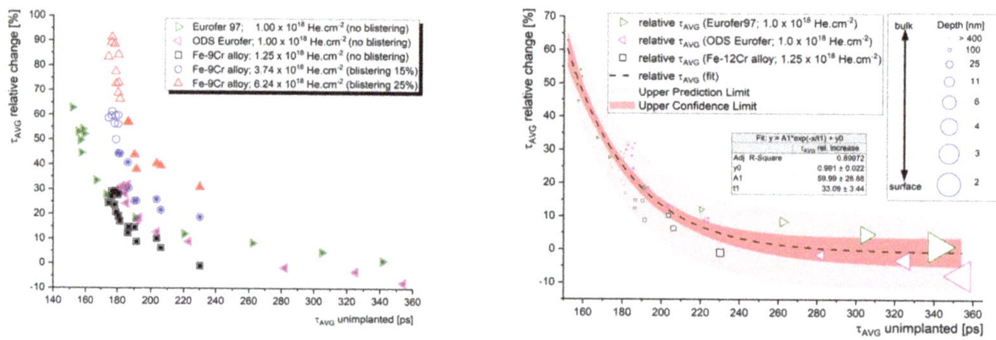

(a) (b)

Figure 10. Relative change in the average positron lifetime in different irradiation experiments plotted as a function of the corresponding lifetime of the unirradiated sample (**a**). Exponential decay fit of datapoints obtained for three different Fe–9Cr alloys irradiated to He fluences $\leq 1.25 \times 10^{18}$ cm^{-2} (**b**).

Based on Figure 10, we can conclude that a comparison of the average positron lifetime in unirradiated materials can provide a useful estimation of the potential radiation tolerance of the material in the swelling incubation phase. In other words, the effectiveness of the initial defect sinks in as-received material can be conveniently predicted using positron lifetime spectroscopy. This technique is also capable of tracking the microstructural evolution in the bubble or void swelling regime, i.e., when the fluence increases to the point where new internal surfaces emerge in the material bulk, such as large cavities or blisters. In such conditions, the interpretation of PAS data can be supported by, and correlated to, TEM.

In addition to the positron annihilation lifetime measurement of helium-implanted metals, valuable complementary information on the chemical composition at the vicinity of the annihilation site can be obtained by DBS. In our earlier work [70], theoretical calculations of ratio profiles for empty and He-filled vacancy clusters were performed to investigate the possibility of direct He detection in radiation-induced open-volume defects, by means of the DBS technique. The comparison of theoretical momentum profiles for monovacancy and nine-vacancy clusters, with and without the presence of contained helium atoms suggests that a characteristic helium-related peak can be expected in the momentum range of 5–7 $\times$ 10^{-2} m_0c. By comparing experimental momentum profiles of the He-implanted Fe–12Cr alloy obtained from the ion track region (low concentration of implanted He ions) and He peak region, we reported experimental evidence of the anticipated He peak. This was later confirmed by Sabelova et al. [71], on annealed Eurofer97 steel irradiated in a spallation neutron source. Although a direct experimental confirmation of a helium signal in the (C)DBS spectra is usually complicated by various factors (mostly the complexity of the material microstructure), the technique is, without any doubts, a very effective tool for the investigation of helium behavior in irradiated structural materials.

In the work of Sharma et al. [72], DBS measurements of He-implanted Ni–Cr alloys have been conducted, along with comprehensive theoretical calculations of positron lifetime characteristics in this type of alloy. These calculations confirm the findings of previous studies on the effect of helium on positron lifetime in vacancy clusters. Similarly to the earlier reported work of Troev et al. [57], the study by Sharma et al. reports a significant decrease in positron lifetime in vacancy clusters (VC), with an increasing number of implanted helium ions. This effect is slightly more pronounced compared to the same quantity of hydrogen atoms. On the other hand, however, this study shows that with increasing the number of implanted ions surrounding the vacancy defects, the S parameter is decreased, which is attributed to the electron momentum distribution of the helium ions. The authors suggest that in the case of He, 1S and 2S electrons have a very large contribution to the low-momentum region (0–1.8 × 10^{-3} m_0c). As a result, the S-parameter value drastically increases with the increase in He atoms surrounding the vacancy defects. This calculation is practically in agreement with the calculations of the characteristic helium peak by Sabelova [70], which indicated that the position of the peak was somewhere in between the regions generally attributed to the S and W parameters. In the experimental part, Sharma et al. reported that the relative density of isolated vacancy defects in the cascade (damage peak) region was smaller compared to the ion track region in the investigated He-irradiated alloys.

Somewhat higher momentum values (8–14 × 10^{-3} m_0c) were attributed to the helium peak in the CDBS spectra of the Fe–9Cr alloy irradiated by He ions at 550 °C, in the work of Zhu et al. [73]. In this work, the authors investigated the effect of irradiation temperature on the formation of helium–vacancy clusters. The reported results suggest a coexistence of large amounts of He_mV_1 and monovacancies in the sample irradiated at room temperature. However, irradiation at a temperature between 250 and 550 °C leads to the absorption of helium atoms by helium–vacancy clusters, and the formation of over-pressurized He_mV_n ($m > n$) clusters or helium bubbles. The results also show that void swelling in the investigated Fe–9Cr alloy is most pronounced under 450 °C irradiation.

Fe–9Cr steel Eurofer 97, together with its ODS variant, implanted by 500 keV helium ions, was investigated using the SPB DBS technique and reported in our recent work [74]. Using VEPFIT analysis, we characterized the positron diffusion in the studied materials and determined the effective positron diffusions length. As expected, this length is significantly shorter in the ODS materials (5 nm in the ODS Eurofer vs. 35 nm in as-received Eurofer97), and the sample preparation plays an important role in the evolution of the radiation-induced point defects in the near-surface region.

The near-surface region of helium-implanted Eurofer97 steel was reported earlier by Carvalho et al. [75]. In this study, extraordinarily low energy of implanted helium ions (30 keV) was used in the irradiation experiments, providing a damage peak at a depth of ~20 nm. Despite the likely strong sinking of the created defects by the sample surface, the results show a clear effect of the initial microstructure on the evolution of helium–vacancy clusters. In the as-received (polished) samples, a decrease in the S parameter was observed after implantation, while no significant change in the S parameter was observed after the implantation of annealed samples. This was explained by helium absorption by pre-existing vacancies and clusters of vacancies, which dominated the creation of new defects. Similar conclusions were obtained for the early stage of the spallation neutron source irradiation of the Fe–9Cr alloy, and published in the paper by Krsjak et al. [37]. The paper by Carvalho et al. further confirms that an irradiation temperature of 100 °C leads to the formation of small helium–vacancy clusters, while large helium bubbles are being formed already at 250 °C. In this regard, the Doppler broadening data are in good agreement with the thermal desorption spectroscopy (TDS) results.

Chinese low-activation martensitic steel CLAM was investigated after 140 keV He implantation by DBS and nanoindentation techniques, in the work by Xin et al. [76]. The work concludes that the concentration of vacancy-type defects decreases with increasing irradiation temperature, ranging from RT to 600 °C. The results of nanoindentation show that the irradiation-induced hardening, observed at all irradiation temperatures, is most pronounced at 200 °C. This suggests that both vacancy–helium complexes and helium bubbles substantially contribute to irradiation-induced hardening.

The behavior of open-volume defects in stainless steel 316L was investigated by the DBS technique, as a function of post-irradiation annealing temperature, by L. Song et al. [77]. The authors reported competition between the formation of helium–vacancy clusters and the recombination of radiation-induced vacancies in the temperature range 100–400 °C. Above 500 °C, a steady-state growth of helium bubbles was observed. After the annealing temperature was increased to above 800 °C, unstable He_nV_m clusters were observed to dissociate into vacancies and He atoms, and eventually disappear. This was observed via a decrease in the S parameter in the near-surface region, as the vacancies sank to the surface and the He atoms diffused out of the sample.

The same stainless steel (316L), and an Fe16.7Cr14.5Ni model alloy, were studied by DBS after irradiation by 140 keV He ions at different fluences, by Lu et al. [78]. In addition to room-temperature irradiation, the investigated samples were irradiated at 300 °C. The authors reported (He-free) vacancy-type defects as a major defect type in the ion track region, while the main defects in the cascade region were identified as He–vacancy complexes. Only small differences were observed between the microstructural changes in the two studied materials after room-temperature irradiation, while irradiation at 300 °C led to more pronounced differences. The much more dramatic increase in the S parameter in the model alloy was explained by the effect of microelements in 316 SS. The authors confirmed previous studies suggesting that impurities, such as C, N, H, and He, can potentially increase the vacancy migration energy in Fe-based alloys.

The irradiation conditions and the used experimental techniques reviewed in this section are summarized in Table 3 below.

Table 3. Selected PAS studies published on helium irradiation of nuclear structural materials.

Material	Ion (E)	T [°C]	Fluence(s) [cm^{-2}]	Techniques	Ref.
Fe–12%Cr alloy	He (250 keV)	<80	1.25×10^{18}	LT (SPB 1 to 18 keV) DBS (SPB 0.5 to 38 keV)	[70]
Ni–40% Cr alloy	He (65 keV)	-	3×10^{16}; 2×10^{17}; 7×10^{17}	DBS (SPB 0.2–20 keV) + VEPFIT analysis; Theoretical calculations	[72]
Fe–9%Cr alloy	He (100 keV)	250; 350; 450; 600	5×10^{16}	PALS (^{22}Na); DBS (SPB ~0.1 to ~20 keV); CDBS (SPB 8 keV)	[73]
Fe–9%Cr–1%W steel (Eurofer97) + its ODS variant	He (500 keV)	RT	1×10^{18}	DBS (SPB 0.5 to 36 keV)	[74]
Fe–9%Cr–1%W steel (Eurofer97)	He (3 keV)	100; 250	3×10^{19} (100 °C) 1–2 $\times 10^{19}$ (250 °C)	DBS (SPB 0.1 to 25 keV); TDS	[75]
Fe–9%Cr steel (CLAM)	He (140 keV)	RT–600	1×10^{16}	DBS (SPB ~0.1 to ~21 keV) + VEPFIT analysis; Nanoindentation	[76]
316L stainless steels	He (50 keV)	RT (post-irr. anneal. 100–1000)	1×10^{16}	DBS (SPB 0.18 to 20 keV)	[77]
FeCrNi alloy 316L stainless steels	He (140 keV)	RT; 300	1×10^{16} (RT); 5×10^{16} (RT; 300 °C)	DBS (SPB 0.18 to 20 keV)	[78]

3.4. *Synergistic Effects of Helium and Hydrogen on Self-Ion-Induced Damage in Steels*

The production of transmutation helium in structural materials exposed to harsh radiation environments rarely occurs without accompanying production/accumulation of hydrogen. Whether it is fusion or spallation application, structural materials suffer from synergistic effects of hydrogen and helium accumulated in the microstructure of the irradiated material. One of the great opportunities to study these synergistic effects is spallation neutron target irradiation. With a helium production of ~75–90 appm/dpa and H production of ~300–400 appm/dpa, spallation neutron target irradiation provides unique irradiation conditions for such a study. On the other hand, the production rate of the transmutation elements is proportional to the proton/neutron flux, which naturally determines the irradiation temperature gradient. It is, therefore, very difficult to investigate any standalone irradiation parameter and its effect on the microstructure in spallation neutron target irradiation.

Numerous studies on the experiments were published recently, using sequential or simultaneous hydrogen, helium, and self-ion implantation in a wide range of fluences and irradiation or post-irradiation annealing temperatures. The convenient control of the irradiation parameters enabled various effects to be studied independently, which provided a unique set of data for improving the knowledge in this challenging area of nuclear material research.

One of the first studies utilizing slow positron beam experiments on sub-sequential H- and He-implanted samples was published on CLAM steel by Xin et al. [79]. In reference to unirradiated material, a positron diffusion length of 120 $\pm$ 8 nm was obtained by VEPFIT analysis. This is a relatively high value for martensitic steel, and suggests a rather low concentration of positron traps in pristine material. The value of the S parameter in the samples pre-implanted by H was found to be higher compared to the samples pre-implanted by He. This difference decreased with increasing fluence.

At about the same time, Kögler et al. investigated open-volume defect generation and its impact on the hardness of ODS FeNiCr in a dual-beam experiment, applying sequential, as well as simultaneous, Fe and He irradiation [80]. The unirradiated materials contained dislocations and clusters of 4–5 vacancies that were closely related to Y-Al oxide nanoparticles (being less pronounced for samples having undergone heat treatment prior to irradiation), whereas, after irradiation, significant differences were found in the hardness increase for the simultaneous and sequential process in the case of the non-heat-treated sample. This effect was attributed to the immediate stabilization of vacancy clusters by helium during simultaneous irradiation, in contrast to the sequential process, where helium implantation followed Fe. The heat-treated sample, on the other hand, did not exhibit significant differences owing to the irradiation sequence, which, however, might have been caused by a He concentration that was too low. At elevated temperatures, 300 °C, simultaneous dual implantation resulted in a surprisingly large decrease in open-volume defects, and the remaining defects were 3–4 vacancy clusters, in connection with the Y-Al oxide particles, similarly to the untreated sample.

Another Chinese reduced activation ferritic/martensitic (RAFM) steel SIMP, with enhanced high-temperature oxidation and liquid metal corrosion resistance, was investigated by DBS and TEM, after sequential H and He (room temperature) implantation, in a very comprehensive work of Jin et al. [81]. Similarly to the study by Xin et al. [79], the increase in the S parameter of the H-implanted sample was found to be higher than in the case of the He-implanted sample, despite the lower displacement damage introduced by hydrogen implantation. The authors interpreted this by the higher trapping rate of helium atoms by vacancies, compared to H atoms. Helium is expected to occupy the center of vacancies. Any sequential implantation by both He and H ions led to a lower ΔS parameter compared to separate H implantation. The increase in the S parameter He + H was more pronounced compared to the H + He sequence when the hydrogen fluence was sufficiently high. The TEM results indicated that a smaller size and higher density of bubbles are produced in the steel after He + H implantation compared to the hydrogen-only implanted sample. The authors, moreover, conclude that the hydrogen atom prefers the interstitial positions in the vicinity of vacancies and tends to form a H–vacancy complex. Hydrogen is, nevertheless, effectively captured by helium–vacancy clusters, and the existence of such agglomerations leads to the suppression of direct binding between H atoms.

The effectivity of helium–vacancy nano-clusters in storing hydrogen atoms was investigated in detail in work by Zhu et al. [82]. The authors used first-principle calculations and positron annihilation spectroscopy to study the potential of defect sites, formed after irradiation, for hydrogen storage. The study confirms, again, the effectivity of hydrogen capturing by helium–vacancy clusters or bubbles formed in He implantation.

A study focusing on hydrogen-assisted embrittlement of reactor pressure vessel (RPV) steels was reported by Shi et al. [83]. The authors used proton and Fe^{13+} ion irradiation to simulate a harsh radiation environment and study relatively (compared to the usual displacement damage in RPV steels) severely damaged samples. The study shows that the S parameter does not saturate for proton-irradiated (A508-3) steel, even at a relatively high displacement damage (2.26 dpa). On the other hand, Fe ion implantation leads to fast S-parameter saturation, and further irradiation does not increase positron trapping at radiation-induced defect agglomerations. It is reasonable to assume that Fe implantation at 3 MeV produces voids that become sinks for new radiation-induced vacancies, and prevents further production of new helium–vacancy agglomerations.

A comprehensive study of RPV steels under proton and ion irradiation has been published by Jiang et al. [84], employing PAS, atom probe tomography (APT), transmission electron microscopy (TEM), and nanoindentation. The authors identified a wide range of defects, from small-size defects, such as vacancies, vacancy–solute complexes, dislocation loops, to large-size vacancy clusters and cavities. The study aimed to investigate the difference between the microstructural evolution of RPV steels irradiated by protons (240 keV) and heavy ions at low temperatures. Although the synergistic effect between the different implanted ions was not investigated, the study reports unique data provided by a combination of many experimental techniques. Similarly to the study by Shi et al. [83], positron trapping at radiation-induced defects was observed to saturate rapidly in Fe ion implantation at a relatively low dose. On the one hand, positron trapping increased without a notable saturation in the samples irradiated by protons. The atom probe did not reveal any solute segregation at low irradiation temperatures. On the other hand, TEM showed the formation of dislocation loops after proton irradiation to 1 dpa, which were responsible for the increase in hardness measured by the nanoindentation technique. The authors suggest that proton irradiation induces the migration of vacancies and their agglomeration into vacancy clusters. These are expected to be stabilized by hydrogen atoms. Fe implantation resulted in a high density of interstitial and vacancy clusters, inducing the formation of dislocation loops and corresponding to the increase in nano-hardness.

Interesting PAS studies were published on different irradiation experiments on Indian reduced activation ferritic/martensitic steel (INRAFM). An isochronal annealing study has been published on INRAFM samples that were individually irradiated with H and He ions, and sequentially irradiated with both H and He ions [85]. Among the single-ion irradiated samples, the hydrogen-irradiated sample showed the presence of hydrogen–vacancy complexes with a high H content in the as-irradiated state, and an increase in the S parameter, due to the release of H from these complexes, at 100 °C. Complete recovery of the microstructure was observed after annealing at 400 °C. On the other hand, three distinct stages of annealing were identified in the helium-only irradiated samples. There was a defect annealing stage, with a decrease in the S parameter from as-irradiated to 300 °C, bubble nucleation stage, with a stable S parameter from 300 °C to 400 °C, and the bubble growth region, characterized by an increase in the S parameter, was observed between 400 °C and 700 °C. It is reasonable to assume that the first stage is affected by the diffusion of helium into existing sinks (open-volume defects), which naturally reduces the rate of positron trapping at these defects. Similar behavior was observed, for instance, in the work of Sabelova et al. [71]. The authors concluded that when sequential H and He implantation is used, the nucleation and growth of helium bubbles, due to isochronal annealing, are independent of the irradiation sequence, and this is not affected by the presence of hydrogen.

Another ion irradiation experiment involving DBS characterization has been published on the Indian INRAFM steel in [86]. The samples were irradiated up to 70 dpa by Fe ions at different temperatures, with and without helium (~700 appm) pre-implantation. The DBS data confirmed that radiation-induced vacancy-type defects can be effectively annealed at high-temperature Fe irradiation. Complete recovery from irradiation-induced vacancy-type defects was observed at 400 °C. The sample pre-implanted by helium also exhibited a decrease in the S parameter with an increasing irradiation temperature up to 400 °C, but this was due to helium absorption by vacancy clusters and an increase in the He/V ratio. At a higher irradiation temperature, the formation and growth of helium bubbles was observed by a notable increase in the S parameter. Again, this observation is in good agreement with other PAS studies of the irradiation/annealing temperature dependencies of positron trapping in samples containing injected or transmutation helium.

A different approach to the study of the synergistic effect of hydrogen and helium was used in the work of Li et al. [87]. In this study, the authors used Ni^+ pre-implantation to produce radiation-induced vacancy clusters in the studied samples of the Fe–9Cr alloy. Such samples were consequently irradiated by H and He ions separately or subsequently. The study showed that the S parameter after sequential He + H implantation was almost identical to the S parameter obtained for the samples irradiated by only He ions. The authors concluded that H atoms implanted into He pre-implanted material can be effectively absorbed by He_nV_m clusters by forming He_nV_mH complexes. The authors confirmed that helium ions implanted into H pre-implanted material might decompose H_nV_m clusters and release hydrogen into the matrix. This can be accompanied by a production of additional vacancy defects.

Excellent work by Scepanovic et al. [88] was published on the Fe–14Cr alloy and its ODS variant, subjected to single Fe^+ or He^+ ion irradiations at different doses and temperatures. All the experimental techniques applied, namely, PALS, DBS, and TEM, confirmed a more pronounced production of radiation-induced defects in the non-ODS alloy. The experiments confirmed the role of Y-rich nanoparticles in acting as vacancy sinks, hindering the growth of helium bubbles. The published results indicate that positron traps after the irradiation of the ODS alloy are not associated with nanoparticles, but rather with the ferritic matrix. Similarly, TEM data suggested that helium bubbles are mostly associated with the ferritic matrix in both materials, although the coarsening of the bubbles was clearly hindered in the ODS alloy.

Also of interest is another study of this group on Fe–15Cr steel, where the influence of external magnetic fields on defect evolution was studied and found to be "not insignificant" [89]. In the presence of a B field, the samples show a lower vacancy density and smaller cluster sizes compared to irradiation in the absence of it, yet the significance of this effect will be investigated in the future.

One of the few PAS studies utilizing simultaneous implantation, triple-ion (Fe^+, He^+, H^+) irradiation has been published by Parente et al., on the Fe–12Cr alloy (including its ODS variant) and two different ODS Fe–14Cr alloys [16]. The reported DBS momentum profiles suggest that the defects induced by this irradiation in the ODS Fe-12Cr and non-ODS Fe12Cr alloys appear to have the same structural characteristics as Cr atoms, as they are the nearest neighbors. The authors pointed out that defects with similar characteristics were also present in the unirradiated materials produced by powder metallurgy. The S-parameter curves obtained on the ODS Fe–14Cr alloy samples, irradiated by the triple beam at 600 °C, were found to be depth independent for positron energies >5 keV. The obtained S-parameter values were slightly lower than the corresponding values for the unirradiated alloy. This was explained by the trapping of H^+ and He^+ at the pre-existent vacancy clusters. The CDB spectra of the irradiated ODS Fe–14Cr samples, normalized to the corresponding unirradiated sample, pointed out the association between the new radiation-induced vacancy clusters and Cr atoms.

The irradiation conditions and the used experimental techniques reviewed in this section are summarized in Table 4 below.

Table 4. Selected PAS studies published on synergistic effects of heavy-ion and H/He irradiation on the microstructure of nuclear structural materials.

Material	Ion (E)	T [°C]	Fluence(s) [cm^{-2}]	Techniques	Ref.
Fe–9%Cr steel (CLAM)	H (80 keV); He (140 keV)	RT	H: 5×10^{15}–5×10^{16} He: 1×10^{15}–1×10^{16}	DBS (SPB 0.1–20 keV) + VEPFIT analysis	[79]
ODS FeCrAl (PM2000)	Fe (2.5 MeV) He (350 keV)	RT, 300	2.74×10^{15}/He 2.94×10^{16}/Fe	LT (^{22}Na) DBS, CDBS (SPB 30 eV–35 keV) + VEPFIT analysis Nanoindentation	[80]
Fe–9%Cr steel (SIMP)	H (80 keV, 260 keV); He (130 keV; 500 keV)	RT	H: 2.78×10^{15}–1×10^{18} He: 7.14×10^{15}–1.8×10^{16}	DBS (SPB 0.18–20 keV) TEM	[81]
Fe–9%Cr steel	H (80 keV,); He (140 keV)	RT	H: 5×10^{16} He: 1×10^{16}	LT (^{22}Na); DBS (SPB 0.18–20 keV) + theoretical calculations	[82]
RPV steel (A508-3)	Fe (3 MeV) H (240 keV)	100	H: 2.5×10^{16}–1.13×10^{18} Fe: 4.55×10^{13}–2.06×10^{15}	DBS (SPB 0.5–26 keV)	[83]
RPV steel (A508-3)	Fe (3 MeV) H (110 + 240 keV)	100	0.05–1.0 dpa;	DBS (SPB 0.5–26 keV) LT (SPB 14 keV); APT; TEM; Nanoindentation	[84]
Fe–9%Cr steel (INRAFM)	Fe (1.1 MeV) He (50 + 90 + 130 keV)	200, 400, 500	Fe (70 dpa) He (700 appm)	DBS (SPB 0.2–22 keV) + VEPFIT analysis	[85]
Fe–9%Cr steel (INRAFM)	H (80 keV) He (130 keV)	RT; Post-irradiation annealing up to 700	H (1500 appm); He (450 appm)	DBS (SPB 0.25–22 keV) incl. VEPFIT analysis	[86]
Fe9Cr alloy	Ni (1 MeV); H (50 keV); He (80 keV)	RT (Ni); RT (Ni + H); RT (Ni) + 450 (H); RT (Ni) + 450 (H) 450 (He)	Fe: 3.4×10^{13} H: 1×10^{16} He: 2×10^{15}	DBS (SPB 0.18 keV–20.18 keV)	[87]
Fe14Cr alloy; Fe14Cr ODS	Fe (1 MeV); He (50 keV);	RT (Fe) 400–450 (He)	Fe: 6.5×10^{15} He: 6.5×10^{15}; 1×10^{16}	DBS (SPB ~0.1 keV–30 keV) CDBS (SPB 7 keV; 18 keV)	[88]
Fe15Cr alloy	He (45 keV) Fe (1 MeV)	RT	He: 4.0×10^{16} Fe: 8.24×10^{15}; 1.12×10^{16}	DBS, CDBS (SPB ~0.2 keV–30 keV)	[89]
Fe12Cr ODS12Cr ODS14Cr ODS14CrWTi	Fe (10 MeV); Fe (4 MeV)/ He (1.6 MeV)/ H (500 keV)	RT (Single beam); 600 (Triple beam)	Fe: 5×10^{15}; Fe: 1.49×10^{16}/ He 1.4×10^{15}/ H 3.55×10^{15}	DBS (SPB 1–30 keV) CDBS (SPB 30 keV)	[16]

4. Summary

Over the past two decades, positron annihilation spectroscopy has proven to be an invaluable characterization tool in various radiation damage studies aimed at the evolution of early-stage microstructures. However, due to the wide variety of materials and irradiation conditions to be studied, and the complexity of data analysis, it is often difficult to reflect the knowledge obtained so far in new irradiation studies. This paper summarizes recent studies of structural materials subjected to different ion beam irradiation experiments and investigated by techniques of positron annihilation spectroscopy. It provides a convenient reference for experiment proposals in the field and enables a quick comparison of the obtained results to the published data. This helps researchers to conduct more reproducible experiments, and will eventually help them to use positron annihilation spectroscopy as a primary characterization tool.

Funding: This research work was supported by the Slovak Research and Development Agency under the contract No. APVV-20-0010, Scientific Grant Agency of the Ministry of Education of Slovak Republic and the Slovak Academy of Sciences VEGA projects No. 1/0382/20 and 1/0395/20 and the European Regional Development Fund, project No. ITMS2014+: 313011W085.

Conflicts of Interest: The authors declare no conflict of interest.

References

1. Was, G.S. *Fundamentals of Radiation Materials Science: Metals and Alloys*, 2nd ed.; Springer: Berlin/Heidelberg, Germany, 2016. [CrossRef]
2. Hoffelner, W. *Materials for Nuclear Plants: From Safe Design to Residual Life Assessments*; Springer-Verlag London Limited: London, UK, 2013. [CrossRef]
3. Phythian, W.J.; English, C.A. Microstructural evolution in reactor pressure vessel steels. *J. Nucl. Mater.* **1993**, *205*, 162–177. [CrossRef]
4. Hautojärvi, P.; Vehanen, A. Introduction to Positron Annihilation. In *Positrons in Solids*; Hautojärvi, P., Ed.; Springer: Berlin/Heidelberg, Germany, 1979; pp. 1–23. [CrossRef]
5. Slugen, V.; Kögel, G.; Sperr, P.; Triftshäuser, W. Positron annihilation studies of neutron irradiated and thermally treated reactor pressure vessel steels. *J. Nucl. Mater.* **2002**, *302*, 89–95. [CrossRef]
6. Davies, L.M. A comparison of Western and Eastern nuclear reactor pressure vessel steels. *Int. J. Press. Vessel. Pip.* **1999**, *76*, 163–208. [CrossRef]
7. Was, G.S.; Petti, D.; Ukai, S.; Zinkle, S. Materials for future nuclear energy systems. *J. Nucl. Mater.* **2019**, *527*, 151837. [CrossRef]
8. Debelle, A.; Barthe, M.F.; Sauvage, T.; Belamhawal, R.; Chelgoum, A.; Desgardin, P.; Labrim, H. Helium behaviour and vacancy defect distribution in helium implanted tungsten. *J. Nucl. Mater.* **2007**, *362*, 181–188. [CrossRef]
9. Lhuillier, P.E.; Belhabib, T.; Desgardin, P.; Courtois, B.; Sauvage, T.; Barthe, M.F.; Thomann, A.L.; Brault, P.; Tessier, Y. Helium retention and early stages of helium-vacancy complexes formation in low energy helium-implanted tungsten. *J. Nucl. Mater.* **2013**, *433*, 305–313. [CrossRef]
10. Slugen, V.; Kuriplach, J.; Ballo, P.; Domonkos, P.; Kögel, G.; Sperr, P.; Egger, W.; Triftshäuser, W.; Domankova, V.M.; Kovac, P.; et al. Positron annihilation Investigations of defects in copper alloys selected for nuclear fusion technology. *Fusion Eng. Des.* **2004**, *70*, 141–153. [CrossRef]
11. Richter, A.; Anwand, W.; Chen, C.L.; Böttger, R. Evaluation of defect formation in helium irradiated Y2O3 doped W-Ti alloys by positron annihilation and nanoindentation. *J. Nucl. Mater.* **2017**, *494*, 294–302. [CrossRef]
12. Labrim, H.; Barthe, M.F.; Desgardin, P.; Sauvage, T.; Blondiaux, G.; Corbel, C.; Piron, J.P. Vacancy defects induced in sintered polished UO 2 disks by helium implantation. *Appl. Surf. Sci.* **2006**, *252*, 3256–3261. [CrossRef]
13. Roudil, D.; Barthe, M.F.; Jégou, C.; Gavazzi, A.; Vella, F. Investigation of defects in actinide-doped UO2 by positron annihilation spectroscopy. *J. Nucl. Mater.* **2012**, *420*, 63–68. [CrossRef]
14. Labrim, H.; Barthe, M.F.; Desgardin, P.; Sauvage, T.; Corbel, C.; Blondiaux, G.; Piron, J.P. Thermal evolution of the vacancy defects distribution in 1 MeV helium implanted sintered UO2. *Nucl. Instrum. Methods Phys. Res. Sect. B Beam Interact. Mater. At.* **2007**, *261*, 883–887. [CrossRef]
15. Xu, Q.; Fukumoto, K.; Ishi, Y.; Kuriyama, Y.; Uesugi, T.; Sato, K.; Mori, Y.; Yoshiie, T. Irradiation damage from low-dose high-energy protons on mechanical properties and positron annihilation lifetimes of Fe-9Cr alloy. *J. Nucl. Mater.* **2016**, *468*, 260–263. [CrossRef]
16. Parente, P.; Leguey, T.; de Castro, V.; Gigl, T.; Reiner, M.; Hugenschmidt, C.; Pareja, R. Characterization of ion-irradiated ODS Fe-Cr alloys by doppler broadening spectroscopy using a positron beam. *J. Nucl. Mater.* **2015**, *464*, 140–146. [CrossRef]
17. Fluss, M.J.; Hosemann, P.; Marian, J. Charged-Particle Irradiation for Neutron Radiation Damage Studies. *Charact. Mater.* **2012**, 1–17. [CrossRef]
18. Krsjak, V.; Dai, Y. Microstructural probing of ferritic/martensitic steels using internal transmutation-based positron source. *J. Nucl. Mater.* **2015**, *465*, 311–315. [CrossRef]
19. Dai, Y.; Brun, R.; Gao, W.; Geissmann, K.; Hahl, S.; Hou, H.; Huang, Y.; Linder, H.P.; Long, B.; Spahr, A.; et al. The fourth SINQ Target Irradiation Program, STIP-IV. *J. Nucl. Mater.* **2012**, *431*, 2–9. [CrossRef]
20. Čížek, J.; Bečvář, F.; Procházka, I. Three-detector setup for positron-lifetime spectroscopy of solids containing 60Co radionuclide. *Nucl. Instrum. Methods Phys. Res. Sect. A Accel. Spectrom. Detect. Assoc. Equip.* **2000**, *450*, 325–337. [CrossRef]
21. Petriska, M.; Slugen, V.; Sabelova, V.; Sojak, S.; Veternikova, J. QtCDB2 software for coincidence Doppler broadening measurement system. *J. Phys. Conf. Ser.* **2013**, *443*, 012086. [CrossRef]
22. Saro, M.; Kršjak, V.; Petriska, M.; Slugeň, V. Sodium-22 source contribution determination in positron annihilation measurements using GEANT4. In Proceedings of the Applied Physics of Condensed Matter (APCOM 2019), Strbske Pleso, Slovakia, 29 July 2019. [CrossRef]

23. Stephanovich, V.A.; Dryzek, J. The exact solution of the diffusion trapping model of defect profiling with variable energy positrons. *Phys. Lett. A* **2013**, *377*, 3038–3047. [CrossRef]
24. Logan, J.V.; Short, M.P.; Webster, P.T.; Morath, C.P. More accurate parameterization of positron implantation depth profiles for the sensitivity range of positron-based characterization techniques. *J. Appl. Phys.* **2020**, *128*, 045105. [CrossRef]
25. Dryzek, J.; Singleton, D. Implantation profile and linear absorption coefficients for positrons injected in solids from radioactive sources ^{22}Na and ^{68}Ge {minus 45 degree rule} 68Ga. *Nucl. Instrum. Methods Phys. Res. Sect. B Beam Interact. Mater. At.* **2006**, *252*, 197–204. [CrossRef]
26. Čížek, J.; Janeček, M.; Vlasák, T.; Smola, B.; Melikhova, O.; Islamgaliev, R.K.; Dobatkin, S.V. The development of vacancies during severe plastic deformation. *Mater. Trans.* **2019**, *60*, 1533–1542. [CrossRef]
27. Dubov, L.Y.; Akmalova, Y.A.; Stepanov, S.V.; Funtikov, Y.V.; Shtotsky, Y.V. Evaluation of positron implantation profiles in various materials for22Na source. *Acta Phys. Pol. A* **2017**, *132*, 1482–1485. [CrossRef]
28. Dressler, R.; Ayranov, M.; Bemmerer, D.; Bunka, M.; Dai, Y.; Lederer, C.; Fallis, J.; Murphy, A.S.; Pignatari, M.; Schumann, D.; et al. Mn samples for nuclear astrophysics: The needs, the possibilities and the sources. *J. Phys. G Nucl. Part. Phys.* **2012**, *39*, 105201. [CrossRef]
29. Dryzek, J.; Horodek, P. GEANT4 simulation of slow positron beam implantation profiles. *Nucl. Instrum. Methods Phys. Res. Sect. B Beam Interact. Mater. At.* **2008**, *266*, 4000–4009. [CrossRef]
30. Saro, M.; Kršjak, V.; Lauko, R.; Slugeň, V. Application of Na-22 positron source to the investigation of ion-implanted iron samples. In Proceedings of the 18th International Conference on Positron Annihilation (ICPA-18), Orlando, FL, USA, 19–24 August 2018. [CrossRef]
31. Everaerts, J.; Salvati, E.; Korsunsky, A.M. Nanoscale Depth Profiling of Residual Stresses Due to Fine Surface Finishing. *Adv. Mater. Interfaces* **2019**, *6*, 1900947. [CrossRef]
32. Horodek, P.; Dryzek, J. GEANT4 simulation of implantation profiles for positrons injected in solids Na and ^{68}Ge/^{68}Ga. *Nukleonika* **2010**, *55*, 21–23.
33. Hugenschmidt, C.; Piochacz, C. NEPOMUC: Neutron induced positron source Munich. *J. Large-Scale Res. Facil. JLSRF* **2015**, *1*, 22. [CrossRef]
34. Schultz, P.J.; Lynn, K.G. Interaction of positron beams with surfaces, thin films, and interfaces. *Rev. Mod. Phys.* **1988**, *60*, 701. [CrossRef]
35. Weisstein, W.E. *Gaussian Function from MathWorld-A Wolfram Web Resource*; CRC Press: Boca Raton, FL, USA, 1999. Available online: https://mathworld.wolfram.com/GaussianFunction.html (accessed on 17 October 2021).
36. Čížek, J. Characterization of lattice defects in metallic materials by positron annihilation spectroscopy: A review. *J. Mater. Sci. Technol.* **2018**, *34*, 577–598. [CrossRef]
37. Krsjak, V.; Kuriplach, J.; Shen, T.; Sabelova, V.; Sato, K.; Dai, Y. Helium behavior in ferritic/martensitic steels irradiated in spallation target. *J. Nucl. Mater.* **2015**, *456*, 382–388. [CrossRef]
38. Bečvář, F.; Čížek, J.; Procházka, I. Performance of a high-resolution digital positron-lifetime spectrometer. *Acta Phys. Pol. A* **2008**, 1279–1284. [CrossRef]
39. Farrell, K.; Maziasz, P.J.; Lee, E.H.; Mansur, L.K. Modification of radiation damage microstructure by helium. *Radiat. Eff.* **1983**, *78*, 277–295. [CrossRef]
40. Eldrup, M.; Jensen, K.O. Positron Trapping Rates into Cavities in Al: Temperature and Size Effects. *Phys. Status Solidi* **1987**, *102*, 145–152. [CrossRef]
41. Van Veen, A.; Schut, H.; de Vries, J.; Hakvoort, R.A.; Ijpma, M.R. Analysis of positron profiling data by means of VEPFIT. In Proceedings of the 4th International Workshop on: Slow-Positron Beam Techniques for Solids and Surfaces, London, ON, Canada, 3–6 July 1990; pp. 171–198. [CrossRef]
42. Krsjak, V.; Hruska, P.; Degmova, J.; Sojak, S.; Noga, P.; Shen, T.; Sabelova, V.; Egger, W.; Slugen, V. A new approach to near-surface positron annihilation analysis of ion irradiated ferritic alloys. *Nanoscale Adv.* **2021**. [CrossRef]
43. Ziegler, J.F.; Ziegler, M.D.; Biersack, J.P. SRIM-The stopping and range of ions in matter (2010). *Nucl. Instrum. Methods Phys. Res. Sect. B Beam Interact. Mater. At.* **2010**, *268*, 1818–1823. [CrossRef]
44. Kaganov, M.I.; Lifshitz, I.M.; Tanatarov, L.V. Relaxation between electrons and the crystalline lattice. *Sov. Phys. JETP* **1957**, *4*, 173–178.
45. Wang, Z.; Jin, Y.; Hou, M.; Jin, G. Modeling of damage creation in metallic materials under swift heavy ion irradiations. *Nucl. Instrum. Methods Phys. Res. Sect. B Beam Interact. Mater. At.* **2000**, *169*, 98–105. [CrossRef]
46. Rutherford, A.M.; Duffy, D.M. Modelling swift heavy ion irradiation in iron. *Nucl. Instrum. Methods Phys. Res. Sect. B Beam Interact. Mater. At.* **2009**, *267*. [CrossRef]
47. Abhaya, S.; Rajaraman, R.; Sarguna, R.M.; Parida, P.K.; David, C.; Amarendra, G. Defect microstructure in high temperature Ni+ implanted FeCrCoNi-a positron beam study. *J. Alloys Compd.* **2019**, *806*, 780–787. [CrossRef]
48. Agarwal, S.; Liedke, M.O.; Jones, A.C.L.; Reed, E.; Kohnert, A.A.; Uberuaga, B.P.; Wang, Y.Q.; Cooper, J.; Kaoumi, D.; Li, N.; et al. A new mechanism for void-cascade interaction from nondestructive depth-resolved atomic-scale measurements of ion irradiation-induced defects in Fe. *Sci. Adv.* **2020**, *6*, eaba8437. [CrossRef]
49. He, C.W.; Barthe, M.F.; Desgardin, P.; Akhmadaliev, S.; Behar, M.; Jomard, F. Positron studies of interaction between yttrium atoms and vacancies in bcc iron with relevance for ODS nanoparticles formation. *J. Nucl. Mater.* **2014**, *455*, 398–401. [CrossRef]

50. Zhu, H.; Wang, Z.; Cui, M.; Li, B.; Gao, X.; Sun, J.; Yao, C.; Wei, K.; Shen, T.; Pang, L.; et al. Temperature dependent surface modification of T91 steel under 3.25 MeV Fe-ion implantation. *Appl. Surf. Sci.* **2015**, *326*, 1–6. [CrossRef]
51. Zhu, H.; Wang, Z.; Gao, X.; Cui, M.; Li, B.; Sun, J.; Yao, C.; Wei, K.; Shen, T.; Pang, L.; et al. Positron annihilation Doppler broadening spectroscopy study on Fe-ion irradiated NHS steel. *Nucl. Instrum. Methods Phys. Res. Sect. B Beam Interact. Mater. At.* **2015**, *344*, 5–10. [CrossRef]
52. Pecko, S.; Heintze, C.; Bergner, F.; Anwand, W.; Slugeň, V. Fe_{2+} ion irradiated JRQ steel investigated by nanoindentation and slow-positron Doppler broadening spectroscopy. *Nucl. Instrum. Methods Phys. Res. Sect. B Beam Interact. Mater. At.* **2018**, *415*, 1–8. [CrossRef]
53. Echenique, P.M.; Flores, F.; Ritchie, R.H. Dynamic screening: Capture and loss processes of protons moving in solids. *Nucl. Instrum. Methods Phys. Res. Sect. B Beam Interact. Mater. At.* **1988**, *33*, 91–97. [CrossRef]
54. Astrelin, V.T.; Burdakov, A.V.; Bykov, P.V.; Ivanov, I.A.; Ivanov, A.A.; Jongen, Y.; Konstantinov, S.G.; Kudryavtsev, A.M.; Kuklin, K.N.; Mekler, K.I.; et al. Blistering of the selected materials irradiated by intense 200 keV proton beam. *J. Nucl. Mater.* **2010**, *396*, 43–48. [CrossRef]
55. Troev, T.; Markovski, A.; Peneva, S.; Yoshiie, T. Positron lifetime calculations of defects in chromium containing hydrogen or helium. *J. Nucl. Mater.* **2006**, *359*, 93–101. [CrossRef]
56. Shivachev, B.L.; Troev, T.; Yoshiie, T. Positron lifetime computations of defects in nickel containing hydrogen or helium. *J. Nucl. Mater.* **2002**, *306*, 105–111. [CrossRef]
57. Troev, T.; Popov, E.; Staikov, P.; Nankov, N. Positron lifetime studies of defects in alpha-Fe containing helium. *Phys. Status Solidi C* **2009**, *6*, 2373–2375. [CrossRef]
58. Horodek, P.; Kulik, M. Application of positron beam for the long range effect studies in proton implanted iron. *Nucl. Instrum. Methods Phys. Res. Sect. B Beam Interact. Mater. At.* **2019**, *443*, 84–89. [CrossRef]
59. Zhang, W.; Shen, Z.; Tang, R.; Jin, S.; Song, Y.; Long, Y.; Wei, Y.; Zhou, X.; Chen, C.; Guo, L. Proton-irradiation induced defects in modified 310S steels characterized with positron annihilation spectroscopy and transmission electron microscopy. *Nucl. Instrum. Methods Phys. Res. Sect. B Beam Interact. Mater. At.* **2018**, *427*, 1–8. [CrossRef]
60. Pecko, S.; Sojak, S.; Slugeň, V. Comparative study of irradiated and hydrogen implantation damaged German RPV steels from PAS point of view. *Appl. Surf. Sci.* **2014**, *312*, 172–175. [CrossRef]
61. Liu, X.; Wang, R.; Jiang, J.; Wu, Y.; Zhang, C.; Ren, A.; Xu, C.; Qian, W. Slow positron beam and nanoindentation study of irradiation-related defects in reactor vessel steels. *J. Nucl. Mater.* **2014**, *451*, 249–254. [CrossRef]
62. Zhao, D.; Li, S.; Wang, X.; Wang, Y.; Liu, F.; Cao, X. Proton irradiation induced defects in T92 steels: An investigation by TEM and positron annihilation spectroscopy. *Nucl. Instrum. Methods Phys. Res. Sect. B Beam Interact. Mater. At.* **2019**, *442*, 59–66. [CrossRef]
63. Ma, Y.; Ran, G.; Chen, N.; Lei, P.; Shen, Q. Investigation of mechanical properties and proton irradiation behaviors of SA-738 Gr.B steel used as reactor containment. *Nucl. Mater. Energy* **2016**, *8*, 18–22. [CrossRef]
64. Sato, K.; Ikemura, K.; Krsjak, V.; Vieh, C.; Brun, R.; Xu, Q.; Yoshiie, T.; Dai, Y. Defect structures of F82H irradiated at SINQ using positron annihilation spectroscopy. *J. Nucl. Mater.* **2016**, *468*, 281–284. [CrossRef]
65. Sato, K.; Xu, Q.; Yoshiie, T.; Dai, Y.; Kikuchi, K. Positron annihilation lifetime measurements of austenitic stainless and ferritic/martensitic steels irradiated in the SINQ target irradiation program. *J. Nucl. Mater.* **2012**, *431*, 52–56. [CrossRef]
66. Das, S. Recent advances in characterising irradiation damage in tungsten for fusion power. *SN Appl. Sci.* **2019**, *1*, 1–20. [CrossRef]
67. Dai, Y.; Odette, G.R.; Yamamoto, T. *The Effects of Helium in Irradiated Structural Alloys*; Elsevier Inc.: Amsterdam, The Netherlands, 2012. [CrossRef]
68. Garner, F.A.; Oliver, B.M.; Greenwood, L.R.; James, M.R.; Ferguson, P.D.; Maloy, S.A.; Sommer, W.F. Determination of helium and hydrogen yield from measurements on pure metals and alloys irradiated by mixed high energy proton and spallation neutron spectra in LANSCE. *J. Nucl. Mater.* **2001**, *296*, 66–82. [CrossRef]
69. Krsjak, V.; Degmova, J.; Sojak, S.; Slugen, V. Effects of displacement damage and helium production rates on the nucleation and growth of helium bubbles–Positron annihilation spectroscopy aspects. *J. Nucl. Mater.* **2018**, *499*, 38–46. [CrossRef]
70. Sabelova, V.; Krsjak, V.; Kuriplach, J.; Petriska, M.; Slugen, V.; Veternikova, J. Simeg, Characterization of helium implanted Fe-Cr alloys by means of positron annihilation methods. *J. Nucl. Mater.* **2014**, *450*, 54–58. [CrossRef]
71. Sabelova, V.; Krsjak, V.; Kuriplach, J.; Dai, Y.; Slugen, V. Coincidence Doppler broadening study of Eurofer 97 irradiated in spallation environment. *J. Nucl. Mater.* **2015**, *458*, 350–354. [CrossRef]
72. Sharma, S.K.; Saini, S.; Srivastava, A.P.; Pujari, P.K. Identification of chemical surrounding and type of vacancy defects in the damaged region of ion irradiated Ni–Cr alloy. *Materialia* **2021**, *15*, 100999. [CrossRef]
73. Zhu, T.; Jin, S.; Zhang, P.; Song, L.; Lian, X.; Fan, P.; Zhang, Q.; Yuan, D.; Wu, H.; Yu, R.; et al. Characterization of helium-vacancy complexes in He-ions implanted Fe9Cr by using positron annihilation spectroscopy. *J. Nucl. Mater.* **2018**, *505*, 69–72. [CrossRef]
74. Degmová, J.; Kršjak, V.; Shen, T.; Veterniková, J.Š.; Gatciová, A.; Sojak, S.; Hruška, P. Near-surface investigation of positron diffusion length in helium-implanted Fe9Cr and its ODS variant. *Appl. Surf. Sci.* **2021**, *538*, 148004. [CrossRef]
75. Carvalho, I.; Schut, H.; Fedorov, A.; Luzginova, N.; Desgardin, P.; Sietsma, J. Helium implanted RAFM steels studied by positron beam Doppler Broadening and Thermal Desorption Spectroscopy. *J. Phys. Conf. Ser.* **2013**, *443*, 12034. [CrossRef]
76. Xin, Y.; Ju, X.; Qiu, J.; Guo, L.; Chen, J.; Yang, Z.; Zhang, P.; Cao, X.; Wang, B. Vacancy-type defects and hardness of helium implanted CLAM steel studied by positron-annihilation spectroscopy and nano-indentation technique. *Fusion Eng. Des.* **2012**, *87*, 432–436. [CrossRef]

77. Song, L.; Wang, B.; Zhu, T.; Gong, Y.; Xu, Q.; Guo, L.; Jin, S.; Zhang, P.; Song, Y.; Cao, X. Evolution of defects with isochronal annealing in helium-irradiated 316L studied by slow positron beam. *Nucl. Instrum. Methods Phys. Res. Sect. B Beam Interact. Mater. At.* **2020**, *467*, 80–85. [CrossRef]
78. Lu, E.; Cao, X.; Jin, S.; Zhang, P.; Zhang, C.; Yang, J.; Wu, Y.; Guo, L.; Wang, B. Investigation of vacancy-type defects in helium irradiated FeCrNi alloy by slow positron beam. *J. Nucl. Mater.* **2015**, *458*, 240–244. [CrossRef]
79. Xin, Y.; Ju, X.; Qiu, J.; Guo, L.; Li, T.; Luo, F.; Zhang, P.; Cao, X.; Wang, B. Vacancy-type defect production in CLAM steel after the co-implantation of He and H ion beams studied by positron-annihilation spectroscopy. *J. Nucl. Mater.* **2013**, *432*, 120–126. [CrossRef]
80. Kögler, R.; Anwand, W.; Richter, A.; Butterling, M.; Ou, X.; Wagner, A.; Chen, C.L. Nanocavity formation and hardness increase by dual ion beam irradiation of oxide dispersion strengthened FeCrAl alloy. *J. Nucl. Mater.* **2012**, *427*, 133–139. [CrossRef]
81. Jin, P.; Shen, T.; Cui, M.; Zhu, Y.; Li, B.; Zhang, T.; Li, J.; Jin, S.; Lu, E.; Cao, X.; et al. Study on vacancy-type defects in SIMP steel induced by separate and sequential H and He ion implantation. *J. Nucl. Mater.* **2019**, *520*, 131–139. [CrossRef]
82. Zhu, T.; Wang, B.; Song, L.; Liu, X.; Song, Y.; Liu, Y.; Zhang, P.; Cao, X.; Xu, Q. Sensitivity of positrons at hydrogen storage sites in FeCr alloy containing vacancy and helium atom. *Int. J. Hydrogen Energy* **2020**, *45*, 15571–15577. [CrossRef]
83. Shi, J.J.; Yang, W.; Zhu, Z.J.; Liu, X.B.; Jiang, J.; Wang, R.S.; Wu, Y.C. Slow positron beam study of highly irradiated RPV steel under proton and ion impact. *Radiat. Phys. Chem.* **2019**, *156*, 199–204. [CrossRef]
84. Jiang, J.; Wu, Y.C.; Liu, X.B.; Wang, R.S.; Nagai, Y.; Inoue, K.; Shimizu, Y.; Toyama, T. Microstructural evolution of RPV steels under proton and ion irradiation studied by positron annihilation spectroscopy. *J. Nucl. Mater.* **2015**, *458*, 326–334. [CrossRef]
85. Ramachandran, R.; Julie, S.; Rajaraman, R.; Govindaraj, R.; David, C.; Amarendra, G. High-temperature radiation damage studies of Reduced Activation Ferritic/Martensitic (RAFM) steel at fusion relevant He/dpa ratio using positron beam based Doppler broadening spectroscopy. *J. Nucl. Mater.* **2021**, *544*, 18–22. [CrossRef]
86. Ramachandran, R.; David, C.; Magudapathy, P.; Rajaraman, R.; Govindaraj, R.; Amarendra, G. Study of defect complexes and their evolution with temperature in hydrogen and helium irradiated RAFM steel using positron annihilation spectroscopy. *Fusion Eng. Des.* **2019**, *142*, 55–62. [CrossRef]
87. Li, L.; Jin, S.; Zhang, P.; Wang, D.; Cao, X.; Guo, L.; Xu, Q.; Li, J.; Zhang, T.; Li, L.; et al. Effect of interaction between H and He on micro-defects in Fe9Cr alloy investigated by slow positron beam. *J. Nucl. Mater.* **2019**, *526*, 151748. [CrossRef]
88. Šćepanović, M.; de Castro, V.; García-Cortés, I.; Sánchez, F.J.; Gigl, T.; Hugenschmidt, C.; Leguey, T. Characterisation of open volume defects in Fe–Cr and ODS Fe–Cr alloys after He+ and Fe+ ion irradiations. *J. Nucl. Mater.* **2020**, *538*, 152230. [CrossRef]
89. García-Cortés, I.; Leguey, T.; Sánchez, F.J.; Maira, A.; Moroño, A.; Muñoz, P.; Scepanovic, M.; Marco, J.F. Study of damage in binary Fe 85 Cr 15 alloys irradiated by ions and the effect of an external magnetic field during irradiation. *J. Nucl. Mater.* **2019**, *517*, 138–147. [CrossRef]

materials

MDPI

Article

Comparison between Subsequent Irradiation and Co-Irradiation into SIMP Steel

Yong Wang [1,*], Tongmin Zhang [2], Qing Liao [3], Junyuan Yang [3], Weigang Gu [1], Yongfei Ren [1], Zheng Jia [1] and Bingsheng Li [3,4,*]

1 China Institute for Radiation Protection, Taiyuan 030006, China; cirpguweigang@163.com (W.G.); renyongfei0603@163.com (Y.R.); Jz15383433932@163.com (Z.J.)
2 Institute of Modern Physics, Chinese Academy of Sciences, Lanzhou 730000, China; ztm@impcas.ac.cn
3 State Key Laboratory for Environment-Friendly Energy Materials, Southwest University of Science and Technology, Mianyang 621010, China; liaoqing@163.com (Q.L.); yangjy2019@sina.com (J.Y.)
4 Engineering Research Center of Biomass Materials, Ministry of Education, Southwest University of Science and Technology, Mianyang 621010, China
* Correspondence: wangyong_cirp@163.com (Y.W.); libingshengmvp@163.com (B.L.)

Abstract: A modern Chinese ferritic/martensitic steel SIMP, is a new perspective nuclear structural material for the spallation target in accelerator driven sub-critical system. In this work, aimed at exploring the radiation resistance properties of this material, we investigate the differences between simultaneous Fe and He ions irradiation and He implantation of SIMP steel pre-irradiated by Fe self-ions. The irradiations were performed at 300 °C. The radiation-induced hardening was evaluated by nano-indentation, while the lattice disorder was investigated by transmission electron microscopy. Clear differences were found in the material microstructure after the two kinds of the ion irradiation performed. Helium cavities were observed in the co-irradiated SIMP steel, but not the case of He implantation with Fe pre-irradiation. In the same time, the size and density of Frank loops were different in the two different irradiation conditions. The reason for the different observed lattice disorders is discussed.

Keywords: martensitic steel; Fe and He irradiation; microstructure

Citation: Wang, Y.; Zhang, T.; Liao, Q.; Yang, J.; Gu, W.; Ren, Y.; Jia, Z.; Li, B. Comparison between Subsequent Irradiation and Co-Irradiation into SIMP Steel. *Materials* **2021**, *14*, 1393. https://doi.org/10.3390/ma14061393

Academic Editor: Scott M. Thompson

Received: 4 February 2021
Accepted: 4 March 2021
Published: 12 March 2021

1. Introduction

Nuclear energy is a leading source of low-carbon electricity in the world. Together with renewables, it provides an essential role in the global energy transition. However, three main concerns to be considered: (1) ensuring a sustainable and responsible use of resources; (2) minimizing the probability and consequences of major nuclear accidents; (3) reducing the production of long-lived radionuclides, e.g., ^{129}I, ^{137}Cs, ^{90}Sr, ^{99}Tc in nuclear waste. In order to address all three aspects, the concept of an accelerator driven sub-critical (ADS) system was proposed in the 1990s, containing spallation neutron source, which enables usf to run a subcritical nuclear reactor and to use an existing spent fuel from PWR reactors. In particular, the decay time of the long lived radionuclide can be reduced by a factor 1000 to time scales that are below 1000 years, resulting in shorter-term hazard and less heat from the waste. The Chinese Academy of Sciences (CAS) performed a strategic project of advanced nuclear fission energy in the future since 2011 [1]. Using a very high power of neutron transmutes the long-lived radionuclide into short-lived medium elements, as well as from less than 1% of the fuel's energy content in light water reactors to above 95% in ADS. Due to its a low melting point, low vapor pressure, high atomic number and high thermal conductivity, lead-bismuth eutectic (LBE) is considered the primary candidate material used in spallation target and coolant. One of the serious problems using LBE is that containment materials, such as ferritic-martensitic (FM) steels, suffer from LBE corrosion [2,3]. Oxidation, dissolved corrosion and liquid metal embrittlement possible occur when FM steels contact with LBE. In order to lower the influence of LBE corrosion, a new type of reduced activation martensitic steel, SIMP steel (Fe-10.7 Cr–1.4 Si–1.2 W–0.5 Mn–0.2 C–0.2 V, wt.%) with 1.4 wt.% Si, was corporately developed by the Institute of Modern Physics, CAS and Institute of Metal Research, CAS. In ADS operation surroundings, SIMP steel suffers from synergistic influence of neutron irradiation, LBE corrosion and high temperature. Therefore, it is vital to assess the property changes of SIMP steel under particle irradiation and LBE corrosion.

The use of ion irradiation as a surrogate for neutron irradiation is a convenient way to assess the radiation resistance of nuclear material, in a reasonable time and without any induced radioactivity. So far, most of reports of ions irradiated SIMP steel using single-beam ions [4–7]. It is well known that a great number of helium atoms are formed by the (n, α) nuclear reaction. We can use Fe ion irradiation to simulate neutron irradiation-induced lattice displacement. The production of transmutation helium in the material can be effectively simulated by He ion implantation [8–10]. Due to the simultaneous occurrence of lattice disorder and helium accumulation, simultaneous dual-beam implantation of Fe and He seems to be an optimal procedure for an experimental simulation of realistic radiation condition. However, there are many reports published in the literature where subsequent irradiations are performed, either with pre-implanted Fe followed by He implantation or with pre-implanted He followed by Fe implantation [11–13]. Wei et al. [11] reported irradiation hardening under He and subsequent Fe ions irradiation in CLAM steel (Fe–8.9 Cr–1.5 W–0.5 Mn–0.2 V–0.1 C, wt.%) at RT. They argued significant irradiation hardening via He ions irradiation to fluences of 0.7–1.0 $\times$ 10^{17} ions/cm^2 at room temperature (RT), and less contribution of the subsequent Fe ions irradiation to a fluence of 5.0 $\times$ 10^{15} ions/cm^2 at RT. It is still an open question whether subsequent irradiation can simulate co-irradiation, which needs elucidate what difference between subsequent irradiation and simultaneous irradiation. In this study, we investigated the microstructure of SIMP steel irradiated by Fe + He simultaneously and Fe + He subsequently at 300 °C. The irradiation temperature is within the operation surroundings of ADS and pressurized water reactor. Dislocation loops are particularly concerned, because clusters of self-interstitial atoms and interstitial-type dislocation loops are of the main microstructural damage features at intermediate temperatures (room temperature to 300 °C). These defects act as obstacles for dislocation movement, resulting in irradiation hardening and the increase in ductile-to-brittle transition temperature (DBTT).

2. Experimental Process

The composition of SIMP steel is given in Table 1. The sample surface was polished by sandpaper down to 4000 grit and then mirror-like polished with 0.1 diamond spray. The sample dimension was 5 $\times$ 3 $\times$ 1 mm^3. Dual-beam irradiation experiment was carried out at the dual-Beam Material Irradiation Facility for Energy Technology (DuET) at the Institute of Advanced Energy, Kyoto University. The experiment facility was described in Ref. [14]. The energy of Fe^{3+} ions was 6.4 MeV provided by a 1.7 MeV tandem accelerator. The energy of He^{+} ions was 1 MeV supplied by a 1 MV Singletron accelerator. In order to provide a uniform distribution of helium atoms in a certain depth range, the helium beams went through three different thicknesses of alumina films. The Fe and He fluences were 1.65 $\times$ 10^{16} ions·cm^{-2} and 4.5 $\times$ 10^{15} ions·cm^{-2}, respectively. The irradiation temperature was well controlled at 300 $\pm$ 5 °C detected by an infrared thermometer. The Monte-Carlo code Stopping and Range of Ions in Matter (SRIM-2008) quick cascade simulations was used to simulate depth profiles of the displacement damage induced by 6.4 MeV Fe^{3+} irradiation and the helium deposition (using the threshold displacement energy of 40 eV and the density of 7.86 g·cm^{-3}) [9]. The simulated result is illustrated in Figure 1a. The nominal displacement damage rate, nominal displacement damage were 3 $\times$ 10^{-4} dpa·s^{-1} and 5 dpa at a depth of 600 nm, respectively. The helium injection ratio was 60 appm He.dpa^{-1} in the corresponding depth. Single-beam irradiation experiment was carried out at 320 kV High-voltage Platform in the Institute of Modern Physics, Chinese Academy of

Sciences. 2.5 MeV Fe^{10+} ions to a fluence of 1×10^{16} ions·cm^{-2} at 300 °C irradiated SIMP steel, and then 110 keV He^{+} ions to a fluence of 5×10^{14} ions·cm^{-2} at 300 °C implanted the Fe pre- irradiated sample. Figure 1b shows that the displacement damage and helium injection ratio were 5 dpa and 60 appm He dpa^{-1} at a depth of 300 nm, respectively; therefore, it gets the same displacement damage and helium concentration as the Fe and He co-irradiation. Note that the displacement contribution by He implantation was only 0.01 dpa, which is far smaller than the elastic collision induced by Fe irradiation. The Fe^{10+} flux was kept at 6×10^{11} cm^{-2} s^{-1} and the He^{+} flux was kept at 3×10^{12} cm^{-2} s^{-1}. The depth profiles of the displacement damage induced by 2.5 MeV Fe^{10+} irradiation and the helium deposition were simulated by SRIM-2008 [15], as shown in Figure 1b.

Table 1. Nominal composition of SIMP steel (wt.%).

Steels	Fe	Cr	Si	W	Mn	C	V	Nb	P
SIMP	Bal	10.7	1.4	1.2	0.5	0.2	0.2	0.01	0.004

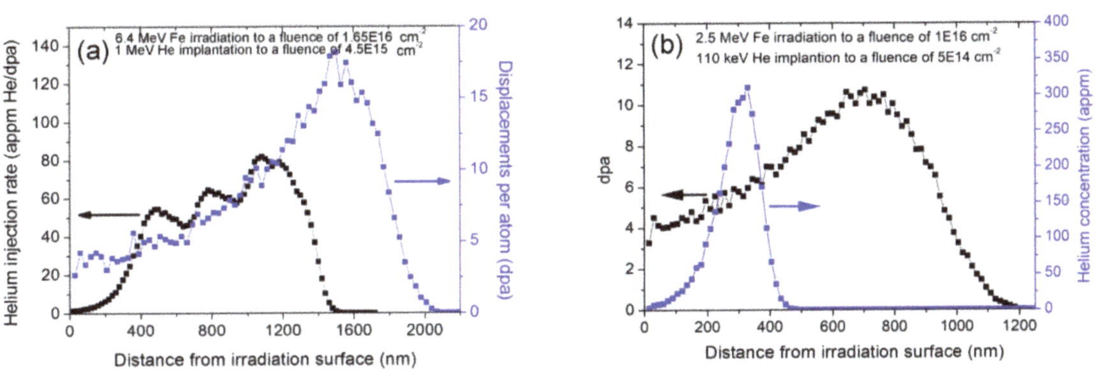

Figure 1. Depth profiles of displacement damage induced by Fe irradiation and of helium deposition via (**a**) dual-beam, (**b**) single-beam, simulated by SRIM-2008.

Nano Indenter G200 with a Berkovich-type indentation tip with a load resolution of 50 nN and displacement resolution of 0.01 nm was used to measure the nano-indentation hardness. The continuous stiffness measurement (CSM) was carried out to characterize the hardness-depth relationship. The maximum penetration depth was 1.2 µm with a strain 0.05 s^{-1} rate. A harmonic displacement of 2 nm, 45 Hz frequency and Poissons Ratio of 0.3 at room temperature, was set. A fused silica reference specimen was used to calibrate the indenter tip geometry. The Oliver–Pharr method was performed to calculate the hardness change. For each sample, fifteen testing points with 50 µm interval were set and the averaged hardness was calculated based on the measured hardness data. Post irradiation, the sample was characterized by a JEOL 2010 TEM operated at 200 kV. Cavities were observed via Fresnel contrast, where they exhibit bright Fresnel fringes in under-focused condition, but dark Fresnel fringes in over-focused condition [10]. Lattice disorder was observed by dynamical two-beam bright-field, as well as weak-beam dark-field diffraction conditions. Cross-sectional TEM (XTEM) samples were fabricated to investigate the depth distribution of radiation-induced damage. XTEM samples were obtained by Hitachi 2000 focused ion beam (FIB) system. Initially 30 keV was used and finally 5 keV was used. Electrochemical polishing for 15 ms at −45 °C was performed to remove the Ga ions-induced damage during FIB fabrication. Initially deposited a 1–2 µm thickness tungsten film on the irradiated surface, TEM-lamella sample was extracted by 30 kV Ga ions. The sample thickness was measured by convergent beam electron diffraction (CBED) method.

The size and number density of the observed cavities were measured by Nano Measurer software with an uncertainty of approximately ±20% [16].

3. Results and Discussion

Figure 2 presents the hardness profile with indentation depth. It can be seen that the hardness value of the unirradiated sample decreases monotonously with increasing indentation depth. It can be accounted for the indentation size effect, which is often observed in crystalline materials and related to the increasing contribution of geometrically necessary dislocations at small scales [17]. Similarly, the hardness values of samples after Fe irradiation and Fe pre-irradiation + He implantation also decrease monotonously with indentation depth. The difference is that the hardness value at depths ranging from 50–700 nm is significantly larger than that of the unirradiated sample. The hardness values between the Fe irradiation and Fe pre-irradiation + He implantation have little difference, which indicates that the He implantation does not introduce pronounce hardness. However, the Fe and He co-irradiated sample has different curve change. There is an inflection at a depth of approximately 300 nm, which is about one fifth of the Fe ion projected range. According to the plastic deformed theory, the depth of the deformed region beneath the indent was approximately 4–5 times indentation depth [18]. This result demonstrates there is an obvious hardening layer inside the sample. Using Nix-Gao model and Kasada approach [19,20] the hardness value in the unirradiated sample is 3.31 GPa, and hardness increased about 43.5%, 44.1–4.75% GPa, 4.77 GPa after Fe irradiation and Fe pre-irradiation + He implantation, respectively. However, the hardness change obviously increases about 50.2–4.97% GPa after Fe and He co-irradiation. Therefore, the Fe and He co-irradiation can enhance hardness of the damaged layer. Lee et al. [21] investigated triple ion beams studies of radiation damage in 9 Cr–2 WVTa at 80, 200 and 350 °C, and reported the severe hardening in Fe + He + H, followed by Fe + He, Fe, Fe + H, He and He + H beams. The present result is consistent with Lee's report. According to the Orowon's theory for athermal bowing of dislocations around obstacles on a slip plane [22], the hardening is proportional to square root of the number density and diameter of obstacles, such as irradiation-induced dislocation loops, cavities and precipitates. Therefore, it is necessary to observe microstructure change in the different irradiated samples.

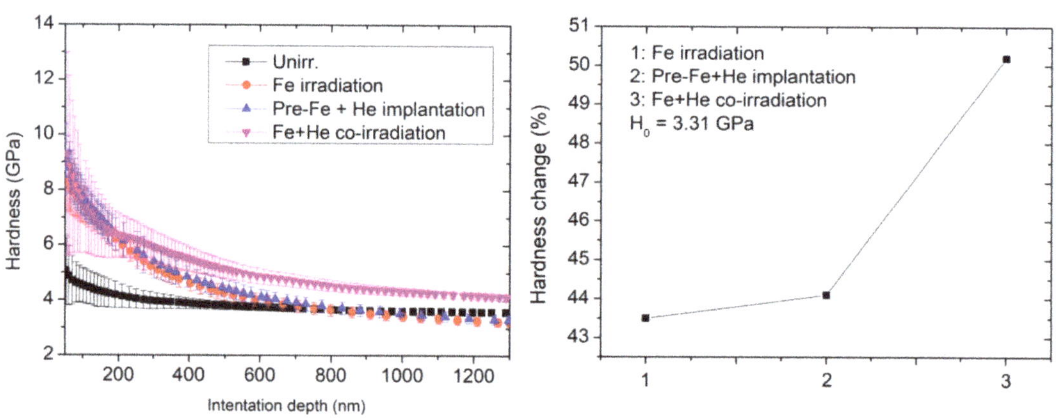

(a) (b)

Figure 2. (**a**) Indentation-depth dependence of the nano-indentation hardness of samples after Fe irradiation, Fe pre-irradiation + He implantation and Fe + He co-irradiation, compared to the as-irradiated sample, (**b**) the hardness change (%) after three different irradiation conditions.

Figure 3 shows the depth distribution from 200 to 700 nm of the observed lattice defects and cavities formed by the Fe and He co-irradiation, where helium deposition was calculated by SRIM-2008. In bright-filed condition, a great number of defects, including black spots, ellipse-shaped loops and lines, are exhibited black contrasts. In general, black spots are regarded as the initial growth stage of ellipse-shaped loops; thus, black spots and ellipse-shaped loops belong to dislocation loops [23]. The size and density of observed dislocation loops are analyzed. In addition, many cavities observed under the under-focused condition, and the distribution zone is consistent with helium deposition simulated by SRIM-2008. No preferential nucleation of cavities along grain boundaries or dislocations is observed. It is indicated that the migration rate of irradiation-induced vacancies is limited at 300 °C, rather than the case of 400 °C and above irradiation that cavities are inclined to form along grain boundaries and dislocations, resulting in increasing risk of brittle-fracture [24].

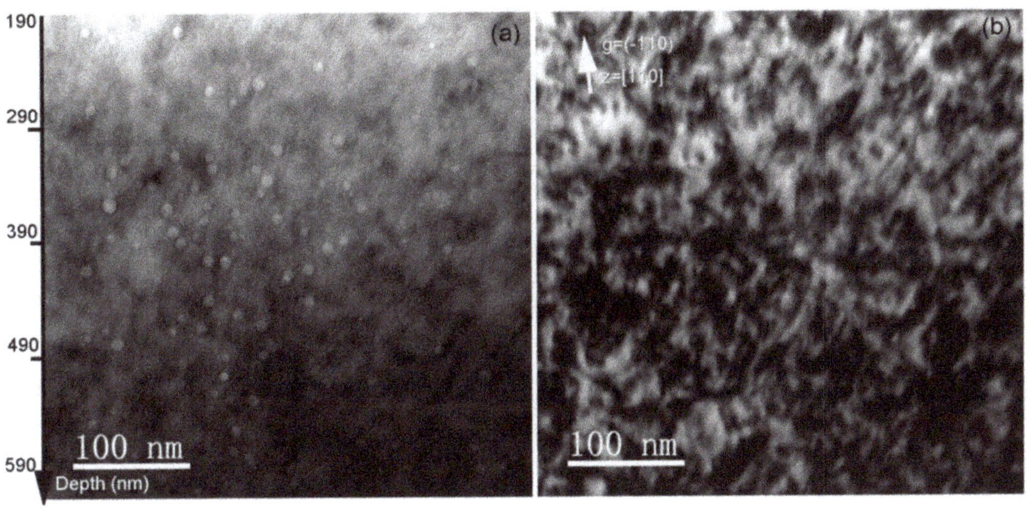

Figure 3. (**a**) Bright-field XTEM micrograph of irradiation-induced dislocation loops in the Fe and He co-irradiated SIMP steel to fluences of 1.65×10^{16} $Fe^{3+}\cdot cm^{-2}$ and 4.5×10^{15} $He^{+}\cdot cm^{-2}$. Micrographs were taken near [110] zone axis under the kinematic condition, $g = (-110)$, (**b**) Bright-field TEM micrograph of cavities formed in the Fe and He co-irradiated SIMP steel. The under-focused view with $\Delta f = -512$ nm was performed to characterize cavity distribution.

Figure 4 shows the depth distribution from 350 to 700 nm of lattice defects, including dislocation loops and dislocation lines in the He implantation with Fe pre-irradiation into SIMP steel. The dislocation loops were formed by Fe irradiation and dislocation lines exist in the as-grown sample. The lattice defects can be observed by the two-beam and weak-beam dark-field conditions. In the two-beam bright-field condition, dislocation loops were easily observed due to a strong diffraction contrast. Figure 4a,b shows the dislocation loops observed by the two-beam condition with diffraction factors, $g = (-1\text{–}10)$ and $g = (200)$, respectively. It can be seen that the number of the observed dislocation loops observed by $g = (200)$ is approximately 3.01×10^{22} m^{-3}, while it is approximately 2.23×10^{22} m^{-3} with $g = (-1\text{–}10)$. The number density of the observed dislocation loops was obtained by the following relationships based on $\boldsymbol{g}\cdot\boldsymbol{b}$ contrast and assuming only 1/2 <111> and <100> loops exist. $d_{111} + 1/3d_{100} = A$ for $g = (200)$, and $1/2d_{111} + 2/3d_{100} = B$ for $g = (110)$, where d_{111} and d_{100} are the number density of 1/2 <111> loops and <100> loops, respectively. A and B are the number density of the visible loops with $g = 200$ and $g = 110$, respectively. Hence, it can get $d_{111} = 2.52 \times 10^{22}$ m^{-3} and $d_{100} = 1.45 \times 10^{22}$ m^{-3}. It demonstrates that the most of the observed dislocation loops have Burgers vectors $\boldsymbol{b} = 1/2$<111>, which is consistent with some reports that 1/2<111> loops were easily formed at 300 °C and below [25,26]. Figure 4d,e shows the lattice defects observed by the weak-beam bright-field and weak-beam dark-field conditions, respectively. Some ellipse-shaped defects and many black spots were observed. The size of the observed lattice defects was evidently smaller than that observed by the two-beam condition, while the number of the observed lattice defects was larger than that observed by the two-beam condition. It should be noted that no cavities were observed in this region.

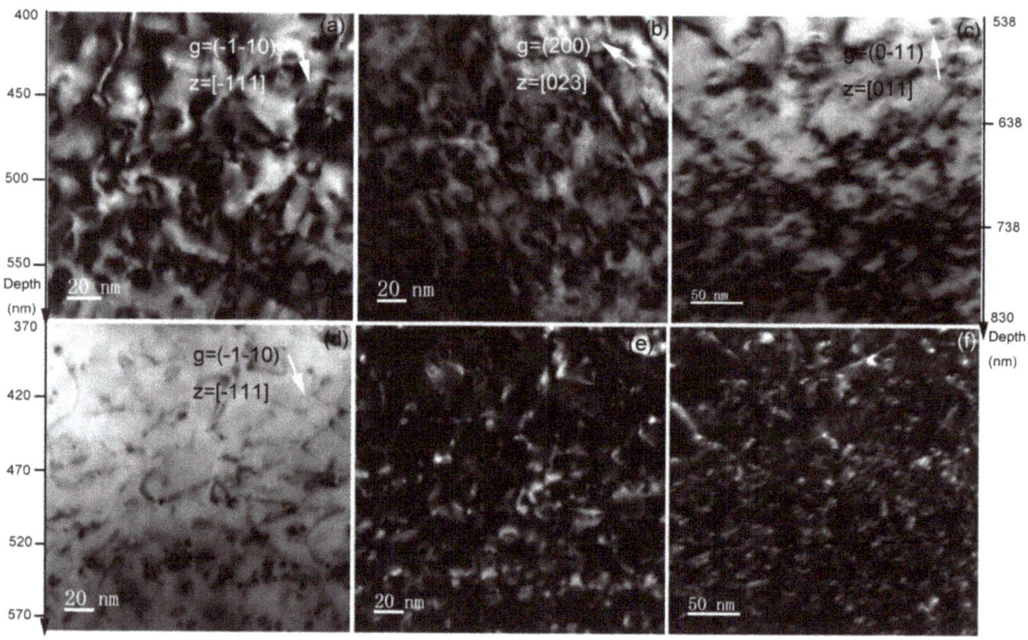

Figure 4. TEM micrographs of irradiation-induced dislocation loops in the He implantation with Fe pre-irradiation into SIMP steel to fluences of 1×10^{16} Fe^{10+}·cm^{-2} and 5×10^{14} He^{+}·cm^{-2}, under bright-field (**a–d**), and weak-beam dark-field (**e**). In comparison, the microstructures observed by bright-field and weak-beam dark field in the Fe-irradiated sample to a fluence of 1×10^{16} Fe^{10+}·cm^{-2} were presented in (**c–f**), respectively.

The size of the observed dislocation loops in Figure 2a was measured and the result is illustrated in Figure 5a. These dislocation loops have size ranging from 4.5–29 nm. The mean size is 12.82 nm and the number density is approximately 1.63×10^{22} m^{-3}. Similarly, the size of the observed cavities in Figure 2b was measured and the result is shown in Figure 5b. The size of these cavities ranges from 2.65–14.35 nm. The mean size is 7.75 nm and the number density is approximately 7.98×10^{21} m^{-3}. In comparison, only dislocation loops were observed in the He implantation with Fe pre-irradiation into SIMP steel, and the observed dislocation loops under the two-beam condition with g = (200) has a mean size of 5.35 nm and a number density of 3.01×10^{22} m^{-3}. According to the present experimental results, there are two differences of the observed lattice defects between the Fe and He co-irradiated SIMP steel and the He implantation with Fe pre-irradiation into SIMP steel. First, cavities were observed clearly in the Fe and He co-irradiated SIMP steel, but not in the He implantation with Fe pre-irradiation into SIMP steel. Tanaka et al. [27] evaluated the synergistic effect via single, dual and triple ion beams consisting of Fe^{3+}, Fe^{3+} + He^{+} or Fe^{3+} + He^{+} + H^{+} in Fe-Cr ferritic alloys, and found the increase in cavity nucleation via dual beam irradiation with He. Wakai et al. [28] argued that swelling in ferritic/martensitic steels is significantly enhanced by the synergistic effect of Fe irradiation- induced damage, hydrogen and helium atoms. Similarly, Hu et al. [29] reported synergistic effect of helium and hydrogen for bubble swelling in ferritic/martensitic steel, and also argued the increasing irradiation swelling of ferritic/martensitic steel after helium and hydrogen implantation. Marian et al. [30] regarded that the synergistic effect of hydrogen-helium depends on irradiation temperature. At lower temperatures (e.g., 470 °C and below), H and He atoms strongly interact with vacancies, but the mobility of these clusters is slow and there are many nano-scaled cavities. It is consistent with our present result that cavities with an average diameter of 8.4 nm were observed in the Fe and He co-irradiated sample. Secondly, the mean size of observed dislocation loops is larger, while the number density is lower after the Fe and He co-irradiation, compared to the He implantation with Fe pre-irradiation. This experimental phenomenon can be attributed to the increasing defect migration due to atomic collision. It is well known that energetic charged particles interact with both electrons and atoms in materials. Kinetic energy transfers to atoms can induce the atom displacement from their original sites, forming interstitial-type defects. The irradiation-induced disorder increases nonlinearly with irradiation dose [31]. With increasing defect concentration, more formed defects start recombination based on radiation-enhanced and diffusion defect reaction rate theory (e.g., high defect concentrations and modified energy barriers due to the formation of charge defects and charge redistribution) [31,32]. Radiation-enhanced diffusion induces the Ostwald ripening of oxide nanoparticles in oxide dispersion strengthened (ODS) steels after Fe irradiation [33]. In addition, the decrease in the threshold temperature for recrystallization during ion collision of amorphous SiC based on the model of ion-beam-induced random nucleation has been reported [34–36], similar phenomenon reported in Si [37]. In addition, the decrease in total lattice disorder after ions implantation or swift-heavy ions irradiation has been observed in disordered SiC [32,38–42]. The same mechanism can explain our experimental results that an obvious increase in the migration rate of lattice defects during Fe and He co-irradiation, resulting in the increase in the mean size, but the decrease in the number density of the observed dislocation loops. Meanwhile, the cavity formation depends on the concentration and size of vacancy-helium clusters. The increase in the migration rate of helium atoms and vacancies induces the formation of vacancy-helium clusters easily, and therefore cavities were observed in the region of helium atom deposition (see Figure 2b). On the contrary, initial Fe irradiation and then He implantation, Fe irradiation-induced defects migrate and accumulate to form clusters, such as vacancies, which are stable, and therefore these defects are hard to migrate during He implantation at a low fluence [32]. The microstructure in the Fe irradiation at 300 °C presents the same morphology with the Fe pre-irradiation + He implantation, as shown in Figure 4c. There are only dislocation loops in the damaged region. Therefore, the influence of He implantation on the Fe pre-irradiation-induced defects can be neglected. As a result, the mean size of the observed dislocation loops is smaller, while the number density of the observed dislocation loops is larger in the He implantation with Fe pre-irradiation into SIMP steel. More important, no cavities are observed in the helium deposition region. The hardness change among these three different irradiations also shows the pronounced irradiation hardening in the Fe and He co-irradiated sample compared to the others. The additional hardening should relate to helium cavities in the damaged layer. It is consistent with report of Ando et al. [43] that extra radiation hardening in the Fe and He co-irradiated F82H due to cavities in the damaged layer. Compared with the co-irradiation, He injection via the subsequent irradiation cannot give effective hardening.

It should be noted that radiation-induced precipitation in nuclear structural materials has been widely reported. Tunes et al. [44] recently reported 30 keV Xe-irradiated an austenitic stainless steel at 800 °C and argued that inert gas bubbles can accelerate clustering and precipitation kinetics. Similarly, vacancy-type defects can enhance the nucleation of carbide precipitations in Ar-irradiated AISI 316L alloys [45]. Fang et al. [46] investigated Fe-irradiated SIMP steel at 300 °C and 400 °C and regarded the Si element enrichment and Ta element depletion inside the precipitates after 1 dpa irradiation at 300 °C. In addition, thermodynamics of the SIMP steel after irradiation should be concerned. Tunes et al. [47]

observed α' and $M_{23}C_6$ precipitates formed in the 30 keV Xe irradiated AISI-348 steel at 800 °C. A equilibrium state can be reached by ion irradiation. Radiation-induced precipitation and radiation-assisted thermodynamic equilibrium can be accounted for the increase in diffusion rate of solute atoms by the exchange of atoms with point defects. In this study, the selected zone diffraction patterns do not present extra diffraction spots, except for matrix, indicating no precipitation formed after irradiation. However, the resolution of the selected zone diffraction pattern is limited. In particular, the growth of carbide precipitates is slow at 300 °C irradiation [48] and, therefore, these new phases have weak diffraction spots as compared to matrix. To confirm irradiation-induced precipitation in SIMP steel, it needs scanning transmission electron microscopy and energy dispersive X-ray spectroscopy to analyze element distribution in the damaged zone.

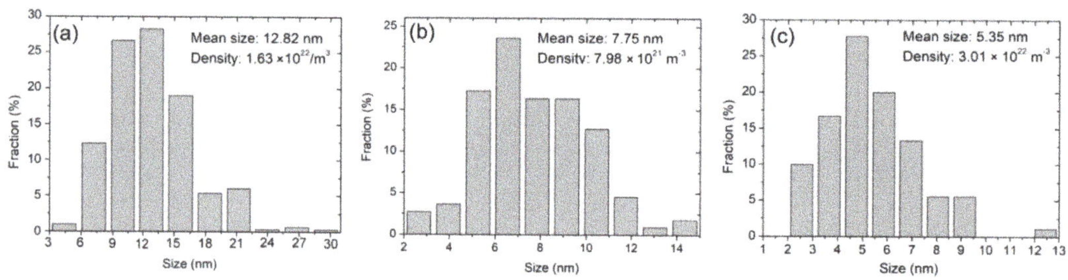

Figure 5. (**a**) Dislocation loop size and (**b**) cavity–diameter distributions in the Fe and He co-irradiated SIMP steel to fluences of 1.65×10^{16} $Fe^{3+}\cdot cm^{-2}$ and 4.5×10^{15} $He^{+}\cdot cm^{-2}$, (**c**) dislocation loop size distribution in the He implantation with Fe pre-irradiation into SIMP steel to fluences of 1×10^{16} $Fe^{10+}\cdot cm^{-2}$ and 5×10^{14} $He^{+}\cdot cm^{-2}$. It shows a skewed distribution in (**a**–**c**).

A schematic diagram presented in Figure 6 illustrates the lattice disorder observed in the two irradiation conditions. There are cavities and dislocation loops formed in the Fe and He co-irradiated SIMP steel at 300 °C, while only dislocation loops were formed in the He implantation with Fe pre-irradiation into SIMP steel. Similarly, no cavities, but only dislocation loops were observed in the Fe irradiated SIMP, as illustrated in Figure 6.

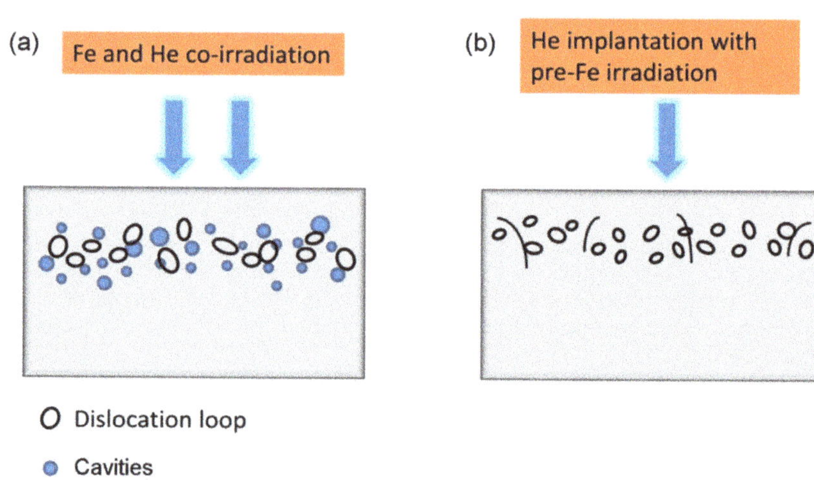

Figure 6. A schematic diagram showing microstructural defects in the Fe and He co-irradiated SIMP steel (**a**) or He implantation with Fe pre-irradiation into SIMP steel at 300 °C (**b**).

4. Conclusions

In this study, a comparison between Fe and He co-implantation and He implantation with Fe pre-irradiation of SIMP steel has been investigated by nano-indenter and XTEM. Irradiation hardening was found more pronounced in the simultaneously co-irradiated samples compared to the subsequently irradiated sample. Almost identical hardness profiles were obtained for the Fe irradiated sample and for the He implanted sample with Fe pre-irradiation. Lattice defects, including dislocation loops and cavities were observed by TEM. By analyzing the observed lattice defects, synergistic effect of simultaneous Fe and He co-irradiation have been observed at 300 °C irradiation, where cavities with a mean size of 7.75 nm are randomly distributed. Moreover, the growth of dislocation loops with a mean size of 12.82 nm was found in the damaged layer, compared to no cavities but small dislocation loops in the subsequent irradiation. Based on the present study, we conclude that subsequent irradiation does not provide an adequate replacement for dual-beam or triple-beam irradiation.

Author Contributions: Original writing & editing, funding acquisition: Y.W.; Irradiation experiment: T.Z.; Investigation: Q.L. and J.Y.; Data curation: W.G.; Formal analysis: Y.R. and Z.J.; Review and Supervision: B.L. All authors have read and agreed to the published version of the manuscript.

Funding: This work is supported by National Nature Science Foundation of China (Grant Nos. U1832133, 12075194) and Sichuan Science Technology Program (Grant No. 2020ZYD055) and National Key Research and Development Program of China (Grant No. 2017YFE0301306).

Institutional Review Board Statement: Not applicable.

Informed Consent Statement: Not applicable.

Data Availability Statement: The data presented in this study are available on request from the corresponding author.

Acknowledgments: This work is supported by National Nature Science Foundation of China (Grant Nos. U1832133, 12075194) and Sichuan Science Technology Program (Grant No. 2020ZYD055) and National Key Research and Development Program of China (Grant No. 2017YFE0301306).

Conflicts of Interest: The authors declare no conflict of interest.

References

1. Zhan, W.L.; Xu, H.S. Advanced Fission Energy Program-ADS Transmutation System. *China Acad. J. Electron. Publ. House* **2012**, *27*, 375–381.
2. Liu, J.; Yan, W.; Sha, W.; Wang, W.; Shan, Y.; Yang, K. Effects of temperature and strain rate on the tensile behaviors of SIMP steel in static lead bismuth eutectic. *J. Nucl. Mater.* **2016**, *473*, 189–196. [CrossRef]
3. Zhang, L.; Yan, W.; Shi, Q.; Li, Y.; Shan, Y.; Yang, K. Silicon enhances high temperature oxidation resistance of SIMP steel at 700 °C. *Corros. Sci.* **2020**, *167*, 108519. [CrossRef]
4. Li, Y.-F.; Shen, T.-L.; Gao, X.; Gao, N.; Yao, C.-F.; Sun, J.-R.; Wei, K.-F.; Li, B.-S.; Zhang, P.; Cao, X.-Z.; et al. Helium-Implantation-Induced Damage in NHS Steel Investigated by Slow-Positron Annihilation Spectroscopy. *Chin. Phys. Lett.* **2014**, *31*, 036101. [CrossRef]
5. Li, Y.-F.; Shen, T.-L.; Gao, X.; Yao, C.-F.; Wei, K.-F.; Sun, J.-R.; Li, B.-S.; Zhu, Y.-B.; Pang, L.-L.; Cui, M.-H.; et al. Cavity Swelling in Three Ferritic-Martensitic Steels Irradiated by 196 MeV Kr Ions. *Chin. Phys. Lett.* **2013**, *30*, 126101. [CrossRef]
6. Wang, J.; Gao, X.; Wang, Z.-G.; Wei, K.-F.; Yao, C.-F.; Cui, M.-H.; Sun, J.-R.; Li, B.-S.; Pang, L.-L.; Zhu, Y.-B.; et al. TEM Characterization of Helium Bubbles in T91 and MNHS Steels Implanted with 200 keV He Ions at Different Temperatures. *Chin. Phys. Lett.* **2015**, *32*, 076101. [CrossRef]
7. Zhu, H.; Wang, Z.; Gao, X.; Cui, M.; Li, B.; Sun, J.; Yao, C.; Wei, K.; Shen, T.; Pang, L.; et al. Positron annihilation Doppler broadening spectroscopy study on Fe-ion irradiated NHS steel. *Nucl. Instrum. Methods Phys. Res. Sect. B Beam Interact. Mater. Atoms* **2015**, *344*, 5–10. [CrossRef]
8. Krsjak, V.; Degmova, J.; Sojak, S.; Slugen, V. Effects of displacement damage and helium production rates on the nucleation and growth of helium bubbles—Positron annihilation spectroscopy aspects. *J. Nucl. Mater.* **2018**, *499*, 38–46. [CrossRef]
9. Degmova, J.; Krsjak, V.; Shen, T.L.; Veternikova, J.S.; Gatciova, A.; Sojak, S.; Hruska, P. Near-surface investigation of positron diffusion length in helium-implanted Fe9Cr and its ODS variant. *Appl. Surf. Sci.* **2021**, *538*, 148004. [CrossRef]
10. Kršjak, V.; Kuriplach, J.; Shen, T.; Sabelová, V.; Sato, K.; Dai, Y. Helium behavior in ferritic/martensitic steels irradiated in spallation target. *J. Nucl. Mater.* **2015**, *456*, 382–388. [CrossRef]

11. Wei, Y.; Liu, P.; Zhu, Y.; Wang, Z.; Wan, F.; Zhan, Q. Evaluation of irradiation hardening and microstructure evolution under the synergistic interaction of He and subsequent Fe ions irradiation in CLAM steel. *J. Alloys Compd.* **2016**, *676*, 481–488. [CrossRef]
12. Yang, Y.; Zhang, C.; Meng, Y.; Liu, J.; Gou, J.; Xian, Y.; Song, Y. Nanoindentation on V–4Ti alloy irradiated by H and He ions. *J. Nucl. Mater.* **2015**, *459*, 1–4. [CrossRef]
13. Jin, P.; Shen, T.L.; Cui, M.H.; Zhu, Y.B.; Li, B.S.; Zhang, T.M.; Li, J.Y.; Jin, S.X.; Lu, E.Y.; Cao, X.Z.; et al. Study on vacancy-type defects in SIMP steel induced by separate and sequential H and He ion implantation. *J. Nucl. Mater.* **2019**, *520*, 131–139. [CrossRef]
14. Kohyama, A.; Katoh, Y.; Ando, M.; Jimbo, K. A new Multiple Beams–Material Interaction Research Facility for radiation damage studies in fusion materials. *Fusion Eng. Des.* **2000**, *51–52*, 789–795. [CrossRef]
15. Interactions of Ions with Matter. Available online: http://www.srim.org (accessed on 12 March 2021).
16. Nano Measurer. Available online: https://www.editsprings.com/Home/ArticleDetail?sort=352394 (accessed on 12 March 2021).
17. Xiao, X.; Yu, L. Nano-indentation of ion-irradiated nuclear structural materials: A review. *Nucl. Mater. Energy* **2020**, *22*, 100721. [CrossRef]
18. Xiao, X.Z.; Chen, L.R.; Yu, L.; Duan, H.L. Modelling nano-indentation of ion-irradiated FCC single crystals by strain-gradient crystal plasticity theory. *Int. J. Plast.* **2019**, *116*, 216–231. [CrossRef]
19. Nix, W.D.; Gao, H. Indentation size effects in crystalline materials: A law for strain gradient plasticity. *J. Mech. Phys. Solids* **1998**, *46*, 411–425. [CrossRef]
20. Kasada, R.; Takayama, Y.; Yabuuchi, K.; Kimura, A. A new approach to evaluate irradiation hardening of ion-irradiated ferritic alloys by nano-indentation techniques. *Fusion Eng. Des.* **2011**, *86*, 2658–2661. [CrossRef]
21. Lee, E.; Hunn, J.; Rao, G.; Klueh, R.; Mansur, L. Triple ion beam studies of radiation damage in 9Cr–2WVTa ferritic/martensitic steel for a high power spallation neutron source. *J. Nucl. Mater.* **1999**, *271–272*, 385–390. [CrossRef]
22. Martin, J.W. *Micromechanisms in Partcle-Hardened Alloys*; Cambridge University: Cambridge, UK, 1980.
23. Jenkins, M.L.; Kirk, M.A. *Characterization of Radiation Damage by Transmission Electron Microscopy*; Informa UK Limited: London, UK, 2001.
24. Li, B.; Wang, Z.; Wei, K.; Shen, T.; Yao, C.; Zhang, H.; Sheng, Y.; Lu, X.; Xiong, A.; Han, W. Evaluation of helium effect on irradiation hardening in F82H, ODS, SIMP and T91 steels by nano-indentation method. *Fusion Eng. Des.* **2019**, *142*, 6–12. [CrossRef]
25. Yao, B.; Edwards, D.; Kurtz, R. TEM characterization of dislocation loops in irradiated bcc Fe-based steels. *J. Nucl. Mater.* **2013**, *434*, 402–410. [CrossRef]
26. Yao, Z.; Hernández-Mayoral, M.; Jenkins, M.; Kirk, M. Heavy-ion irradiations of Fe and Fe–Cr model alloys Part 1: Damage evolution in thin-foils at lower doses. *Philos. Mag.* **2008**, *88*, 2851–2880. [CrossRef]
27. Tanaka, T.; Oka, K.; Ohnuki, S.; Yamashita, S.; Suda, T.; Watanabe, S.; Wakai, E.; Tanaka, T.; Oka, K.; Ohnuki, S.; et al. Synergistic effect of helium and hydrogen for defect evolution under multi-ion irradiation of Fe–Cr ferritic alloys. *J. Nucl. Mater.* **2004**, *329-333*, 294–298. [CrossRef]
28. Wakai, E.; Sawai, T.; Furuya, K.; Naito, A.; Aruga, T.; Kikuchi, K.; Yamashita, S.; Ohnuki, S.; Yamamoto, S.; Naramoto, H.; et al. Effect of triple ion beams in ferritic/martensitic steel on swelling behavior. *J. Nucl. Mater.* **2002**, *307–311*, 278–282. [CrossRef]
29. Hu, W.; Guo, L.; Chen, J.; Luo, F.; Li, T.; Ren, Y.; Suo, J.; Yang, F. Synergistic effect of helium and hydrogen for bubble swelling in reduced-activation ferritic/martensitic steel under sequential helium and hydrogen irradiation at different temperatures. *Fusion Eng. Des.* **2014**, *89*, 324–328. [CrossRef]
30. Marian, J.; Hoang, T.; Fluss, M.; Hsiung, L.L. A review of helium–hydrogen synergistic effects in radiation damage observed in fusion energy steels and an interaction model to guide future understanding. *J. Nucl. Mater.* **2015**, *462*, 409–421. [CrossRef]
31. Was, G.S. *Fundamentals of Radiation Materials Science*; Springer International Publishing: New York, NY, USA, 2017.
32. Zhang, Y.; Weber, W.J. Ion irradiation and modification: The role of coupled electronic and nuclear energy dissipation and subsequent nonequilibrium processes in materials. *Appl. Phys. Rev.* **2020**, *7*, 041307. [CrossRef]
33. Lescoat, M.-L.; Ribis, J.; Chen, Y.; Marquis, E.; Bordas, E.; Trocellier, P.; Serruys, Y.; Gentils, A.; Kaïtasov, O.; de Carlan, Y.; et al. Radiation-induced Ostwald ripening in oxide dispersion strengthened ferritic steels irradiated at high ion dose. *Acta Mater.* **2014**, *78*, 328–340. [CrossRef]
34. Höfgen, A.; Heera, V.; Mücklich, A.; Eichhorn, F.; Skorupa, W. Ion-beam-induced crystal grain nucleation in amorphous silicon carbide. *Nucl. Instrum. Methods Phys. Res. Sect. B Beam Interact. Mater. Atoms* **2000**, *161–163*, 917–921. [CrossRef]
35. Heera, V.; Stoemenos, J.; Kogler, R.; Skurupa, W. Amorphization and recrystallization of 6H-SiC by ion-beam irradiation. *J. Appl. Phys.* **1995**, *77*, 2999–3009. [CrossRef]
36. Heera, V.; Stoemenos, J.; Kögler, R.; Skorupa, W. Complete recrystallization of amorphous silicon carbide layers by ion irradiation. *Appl. Phys. Lett.* **1995**, *67*, 1999–2001. [CrossRef]
37. Nakata, J. Epitaxial crystallization during 600 °C furnace annealing of amorphous Si layer deposited by low-pressure chemical-vapor-deposition and irradiated with 1-MeV Xe ions. *J. Appl. Phys.* **1997**, *82*, 5446–5459. [CrossRef]
38. Han, Y.; Li, B.S.; Wang, Z.G.; Peng, J.X.; Sun, J.R.; Wei, K.F.; Yao, C.-F.; Gao, N.; Gao, X.; Pang, L.-L.; et al. H-ion-induced annealing in He-ion implanted 4H-SiC. *Chin. Phys. Lett.* **2017**, *34*, 12801–12804. [CrossRef]
39. Heinisch, H.; Gao, F.; Kurtz, R. Atomic-scale modeling of interactions of helium, vacancies and helium–vacancy clusters with screw dislocations in alpha-iron. *Philos. Mag.* **2010**, *90*, 885–895. [CrossRef]

40. Wang, X.; Li, J.; Wang, J.; Song, J.; Zhao, F.; Tang, H.; Li, B.; Xiong, A. Microstructure investigation of damage recovery in SiC by swift heavy ion irradiation. *Mater. Des. Process. Commun.* **2019**, *1*. [CrossRef]
41. Thomé, L.; Velişa, G.; Miro, S.; Debelle, A.; Garrido, F.; Sattonnay, G.; Mylonas, S.; Trocellier, P.; Serruys, Y. Recovery effects due to the interaction between nuclear and electronic energy losses in SiC irradiated with a dual-ion beam. *J. Appl. Phys.* **2015**, *117*, 105901. [CrossRef]
42. Debelle, A.; Backman, M.; Thomé, L.; Weber, W.J.; Toulemonde, M.; Mylonas, S.; Boulle, A.; Pakarinen, O.H.; Juslin, N.; Djurabekova, F.; et al. Combined experimental and computational study of the recrystallization process induced by electronic interactions of swift heavy ions with silicon carbide crystals. *Phys. Rev. B* **2012**, *86*, 100102. [CrossRef]
43. Ando, M.; Tanigawa, H.; Jitsukawa, S.; Sawai, T.; Katoh, Y. Evaluation of hardening behavior of ion irradiated reduced activation ferritic/martensitic steels by an ultra-micro-indentation technique. *J. Nucl. Mater.* **2002**, *307–311*, 260–265. [CrossRef]
44. Tunes, M.A.; Schon, C.G.; Greaves, G. Radiation-induced precipitation with concurrent bubbles formation in an austenitic stainless steel (AISI-348). *Materials* **2019**, *7*, 100408–100412.
45. Timm, M.M.; Ítalo, M.O.; Tatsch, F.; Amaral, L.; Fichtner, P.F. Au and Ag ion irradiation effects on the carbide precipitation and Ar bubble formation in solubilized AISI 316L alloys. *Nucl. Instrum. Methods Phys. Res. Sect. B Beam Interact. Mater. Atoms* **2019**, *458*, 174–178. [CrossRef]
46. Fang, X.-S.; Shen, T.-L.; Cui, M.-H.; Jin, P.; Li, B.-S.; Zhu, Y.-B.; Wang, Z.-G. Characterization of Microstructure and Stability of Precipitation in SIMP Steel Irradiated with Energetic Fe Ions *. *Chin. Phys. Lett.* **2017**, *34*, 116102. [CrossRef]
47. Tunes, M.A.; Greaves, G.; Kremmer, T.M.; Vishnyakov, V.M.; Edmondson, P.D.; Donnelly, S.E.; Pogatscher, S.; Schön, C.G. Thermodynamics of an austenitic stainless steel (AISI-348) under in situ TEM heavy ion irradiation. *Acta Mater.* **2019**, *179*, 360–371. [CrossRef]
48. Pareige, C.; Kuksenko, V.; Pareige, P. Behaviour of P, Si, Ni impurities and Cr in self ion irradiated Fe-Cr alloy-Comparison to neutron irradiation. *J. Nucl. Mater.* **2015**, *456*, 471–476. [CrossRef]

materials

MDPI

Article

Bubble Swelling in Ferritic/Martensitic Steels Exposed to Radiation Environment with High Production Rate of Helium

Stanislav Sojak [1,2,*], Jarmila Degmova [1,2], Pavol Noga [2], Vladimir Krsjak [1,2], Vladimir Slugen [1] and Tielong Shen [3]

1 Institute of Nuclear and Physical Engineering, Faculty of Electrical Engineering and Information Technology, Slovak University of Technology in Bratislava, Ilkovičova 3, 81219 Bratislava, Slovakia; jarmila.degmova@stuba.sk (J.D.); vladimir.krsjak@stuba.sk (V.K.); vladimir.slugen@stuba.sk (V.S.)
2 Advanced Technologies Research Institute, Faculty of Materials Science and Technology in Trnava, Slovak University of Technology in Bratislava, Jána Bottu 25, 91724 Trnava, Slovakia; pavol.noga@stuba.sk
3 Institute of Modern Physics Lanzhou, Chinese Academy of Sciences, Lanzhou 730000, China; shentielong@impcas.ac.cn
* Correspondence: stanislav.sojak@stuba.sk

Abstract: Reduced-activativon ferritic/martensitic (RAFM) steels are prospective structural materials for fission/fusion nuclear applications because their radiation and swelling resistance outperforms their austenitic counterparts. In radiation environments with a high production rate of helium, such as fusion or spallation applications, these materials suffer from non-negligible swelling due to the inhibited recombination between vacancy and interstitial-type defects. In this work, swelling in helium-implanted Eurofer 97 steel is investigated with a focus on helium production rates in a wide range of helium/dpa ratios. The results show virtually no swelling incubation period preceding a steady-state swelling of about 2×10^{-4}%/He-appm/dpa. A saturation of swelling above 5000 He-appm/dpa was observed and attributed to helium bubbles becoming the dominant sinks for new vacancies and helium atoms. Despite a relatively low irradiation temperature (65 ± 5 °C) and a rather high concentration of helium, transmission electron microscope (TEM) results confirmed a microstructure typical of ferritic/martensitic steels exposed to radiation environments with high production rates of helium.

Keywords: RAFM steels; swelling; helium implantation; helium bubbles; helium embrittlement

Citation: Sojak, S.; Degmova, J.; Noga, P.; Krsjak, V.; Slugen, V.; Shen, T. Bubble Swelling in Ferritic/Martensitic Steels Exposed toRadiation Environment with High Production Rate of Helium. *Materials* **2021**, *14*, 2997. https://doi.org/10.3390/ma14112997

Academic Editor: Bicai Pan

Received: 22 April 2021
Accepted: 29 May 2021
Published: 1 June 2021

Publisher's Note: MDPI stays neutral with regard to jurisdictional claims in published maps and institutional affiliations.

1. Introduction

Volumetric void, or bubble swelling, is one of the major degradation mechanisms of materials exposed to harsh radiation conditions. It is a key factor limiting the safe operational lifetime of nuclear power plants (NPP) and other nuclear installations. Currently, austenitic steels are used as structural materials at NPPs but their low swelling resistance to neutron irradiations limits their application in environments with high temperatures and dpa rates, which is expected for future nuclear facilities. Similar behaviour can also be observed at irradiation by charged particles [1].

Among structural materials, reduced-activation ferritic/martensitic (RAFM) steels are known for their high resistance to void swelling. Due to their additional low thermal expansion and high thermal conductivity, they are considered as candidate materials for applications in nuclear fusion reactors. RAFM steel Eurofer 97 was chosen as the main blanket structural material for the experimental fusion reactor DEMO.

Substantial research activities are currently underway to develop advanced steels with higher radiation resistance. Physical and mechanical properties of new fabricated model alloys are being tested in harsh radiation environments, such as in spallation neutron targets [2]. The main goal of these research studies is to increase the radiation damage resistance of nuclear structural materials, leading to a long operational lifetime and reliability of fusion reactors [3].

Even though the void swelling resistance of ferritic/martensitic (F/M) steels is fairly adequate, their resistance to bubble swelling after exposure to radiation with high production rates of helium (fusion or spallation) is still questionable. A study on the bubble swelling of 9% Cr and 12% Cr RAFM steel after irradiation to 400 dpa by Fe^{++} at 460 °C, was published by Wang et al. [4]. Results reported 20–300 times lower swelling for 12% Cr steel after irradiation.

This decrease was dependent on the presence of sinks (grain/lath boundaries), oxides, and radiation-induced voids which influence the swelling rate [5,6]. The harsh environment of deuterium–tritium (DT) fusion [3] will likely lead to a very different swelling response compared to that of a fission reactor. Therefore, in-depth microstructural studies are needed to address the swelling phenomenon to improve the radiation resistance of materials to be used in future fusion reactors.

The objective of the study is twofold: to assess the potential of implanting a single-beam ion (helium) to simulate spallation and spallation-relevant radiation environments and to provide a better understanding of the sole effect of the helium production rate on volumetric bubble swelling in F/M steels. To achieve these objectives, we used a low temperature (65 ± 5 °C) helium injection and cross-sectional transmission electron microscope (TEM) analysis. The results are discussed with respect to the helium-to-dpa (c_{He}/dpa) ratio to investigate the performance of F/M steels in a wide range of harsh radiation environments.

2. Experimental

In the present experiments, the RAFM steel "Eurofer 97" (Fe-8.9Cr-1.1W-0.47Mn-0.2V-0.14Ta-0.11C), was investigated. A 10 × 10 × 0.5 mm sample was prepared by electrical-discharge cutting and mechanical polishing. To eliminate any residual stresses induced by the polishing, the surface was given an electrochemical polishing in a perchloric acid solution. The sample was subsequently irradiated at a temperature of 65 ± 5 °C with 500 keV He+ ions with a fluence of 10^{18} ions/cm^2 using the 500 kV open-air implanter at ATRI MTF STU [7]. The number of displacements per atom (dpa), calculated according to the NRT model [8] ranging from 1.6 to 39.4 for the near-surface region (50 nm) and peak region (1050 nm), respectively. The helium concentration (c_{He}) determined in the same way was 370 and 50 atomic parts per million (appm) at.%, respectively. The range of c_{He}/dpa ratios studied in this work was from 460 to 15,000 He-appm/dpa. For a comparison, typical helium production rates in FeCr steel are ~10 He-appm/dpa under fusion reactor conditions and up to 100 appm/dpa in spallation neutron targets [9].

The cross-sectional TEM samples of the implanted steel were prepared using the focused ion beam (FIB) lift-out technique using a FEI Helios Nanolab system (Nanolab Technologies, Milpitas, CA, USA), with Ga acceleration voltage between 1 kV to 30 kV. In the next step, a cleaning process to remove the FIB damaged layer was performed by low energy (Ga) ion milling. Microstructural characterization was performed using FEI Tecnai F20 TEM (FEI Company—Thermo Fisher Scientific, Hillsboro, OR, USA), operated at an accelerating voltage of 200 kV with a field emission gun.

The dpa profile and injected helium concentration profiles, normalized to the area under the curves, are shown in Figure 1 with a cross-sectional TEM micrograph of the implanted steel in the background.

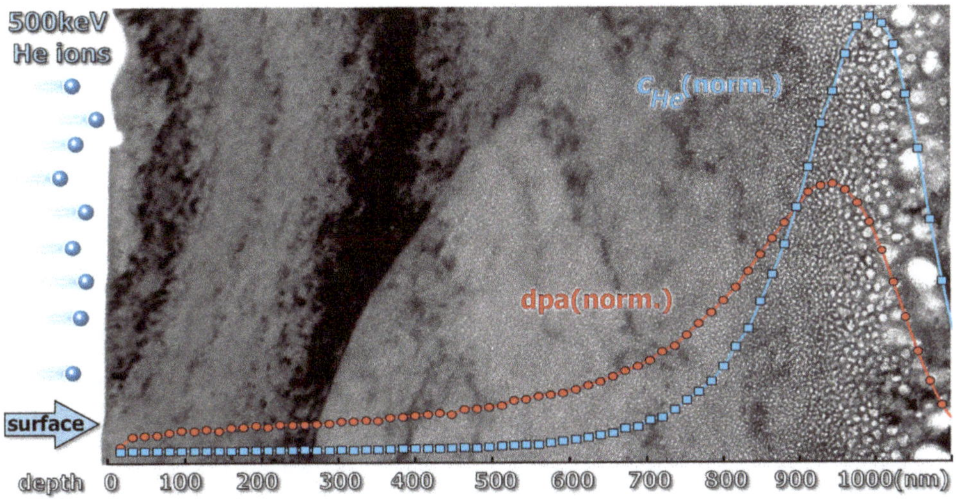

Figure 1. Theoretical simulation of the dpa and c_{He} profiles with a cross-sectional transmission electron microscope (TEM) micrograph of the implanted Eurofer 97 steel in the background.

3. Results

Previously published TEM studies of Eurofer 97 steel [10,11] identified martensite laths of a microns to about 20 µm, with present carbide precipitates along the grain boundaries ($M_{23}C_6$) and inside the martensite grains (MX). Our TEM observation in Figure 2 shows a bright field cross-sectional micrograph of the He-implanted Eurofer 97 steel. As can be seen in the figure, the radiation-induced defects start to be resolvable at a depth of >300 nm, i.e., at ~3 dpa, or ~2000 He-appm, respectively. Two magnified insets, dark field (a) and bright field (b), shown in Figure 2, suggest a relatively uniform microstructure across the mapped area. This area represents very different implantation conditions for helium concentration (c_{He}) and displacement damage (dpa). This implicitly points to the role of the irradiation temperature, a dominant factor in swelling, caused by neutron (produced either in fission or spallation reactions) irradiation in structural steels. Since the temperature is basically constant across the implantation profile, a 200–300% increase in c_{He} and dpa does not lead to a very significant change in microstructure because the F/M steels do not swell at such low displacement damage (~3 dpa) even at high temperatures (>350 °C) [12]. Therefore, the main driving force for volumetric growth is associated with interstitial helium. Another perspective can be obtained from using the ratio of c_{He} and dpa parameters, which change only slightly in the given area. This suggests that the (c_{He}/dpa) parameter is a better indicator (predictor) of volumetric bubble swelling than either displacement damage (dpa) or helium concentration (c_{He}).

The distribution of bubble size corresponding to the region with He concentration 3000–10,000 appm is shown in Figure 3. With average size of ~2 nm the microstructure can be compared to similar F/M steels (F82H) irradiated in a spallation neutron target to ~20 dpa at an irradiation temperature ~300 °C and with a transmutation helium concentration of ~1800 appm [13]. It is clear from the cross-sectional TEM image that the size of bubbles grew significantly (>5 nm) towards the dpa and c_{He} peak.

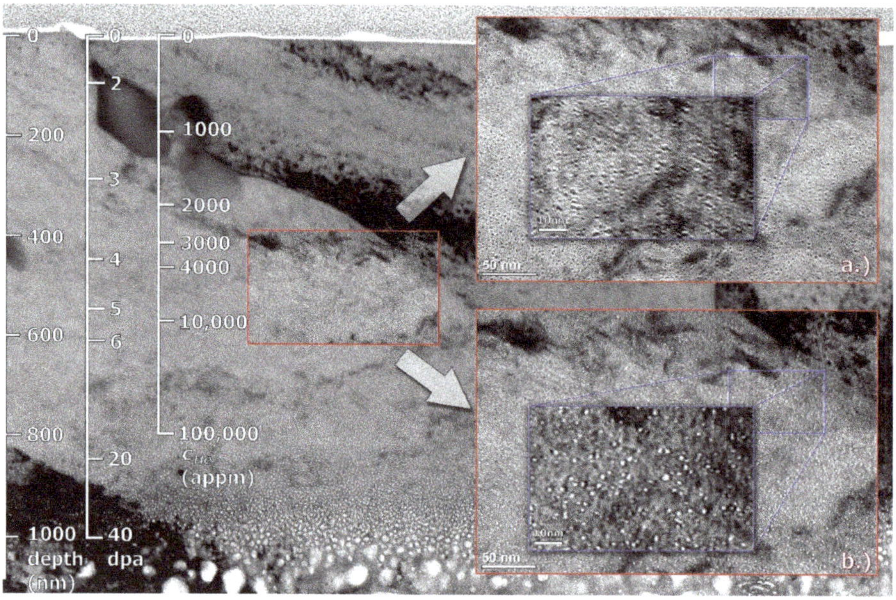

Figure 2. Cross-sectional bright-field TEM micrograph of the implanted steel with magnified dark-field (**a**) and bright-field (**b**) insets representing the range of 3000–10,000 He-appm.

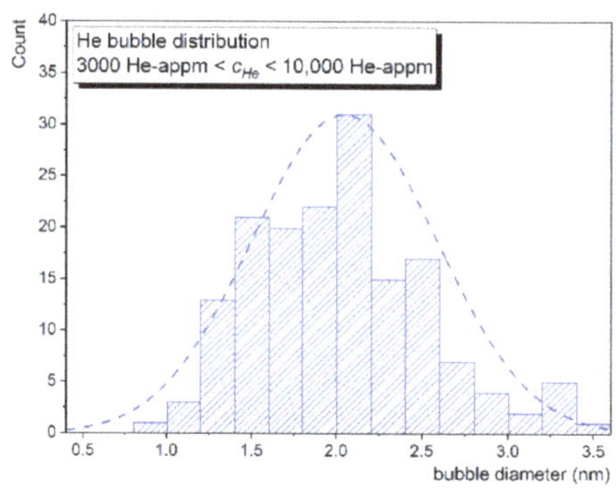

Figure 3. Size distributions of helium bubbles corresponding to c_{He} range 3000–10,000 appm.

In total, 11 areas of at least 50 × 50 nm were analyzed at a depth ranging from 350 to 900 nm, representing the range of 460–15,000 He-appm/dpa. Since the main parameter affecting low-temperature bubble swelling in helium-implanted ferritic steels is the concentration of helium c_{He}, rather than the dpa, it is reasonable to evaluate the swelling rate with respect to the c_{He} or c_{He}/dpa ratio, respectively. Figure 4 shows the volumetric bubble swelling associated with the helium production rate. Since only a weak dependence of the swelling rate on the dpa was expected, it was evaluated only as an average value, i.e., 0.08 ± 0.03%/dpa. This was, in the first approach, surprisingly low with respect to a commonly accepted steady state swelling rate of 0.2%/dpa for F/M steels irradiated under fast reactor conditions [14]. On the other hand, the experimental swelling data in Figure 4 show virtually no transient regime (incubation period) prior to the onset of steady-state swelling, which can be clearly seen between 750 and 3000 He-appm/dpa. The incubation period is subsequently represented by the initial two data points in the figure. The experimental swelling data were saturated after approximately 3000 He-appm/dpa.

This seemed to represent the irradiation conditions where newly formed bubbles were becoming dominant sinks for radiation-induced vacancies and helium. Such a regime would, in material engineering, correspond to massive damage, likely compromising the structural integrity of the given component. Nevertheless, considering the general trend in the plotted data, volumetric bubble swelling in the studied steel could be described as a nonlinear saturation using either logarithmic or a 1-exp function. A more comprehensive assessment of helium bubble evolution in a wide range of He concentrations, that includes complementary positron annihilation experiments and different types of Fe–Cr alloys, is being carried out.

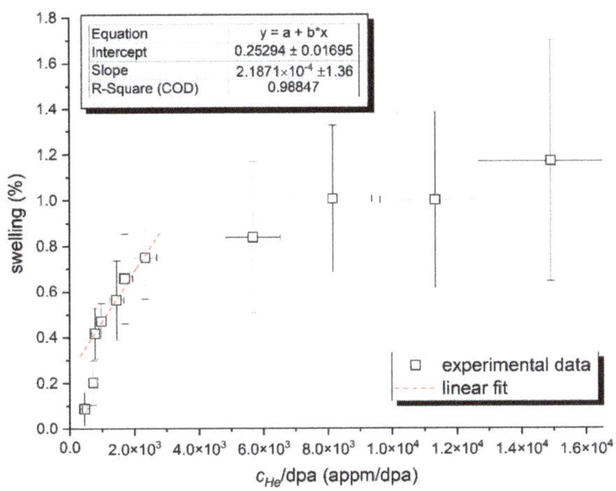

Figure 4. Volumetric bubble swelling in Eurofer 97 steel as a function of helium concentration relative to the number of displacements per atom (c_{He}/dpa).

The increasing dominance of helium bubbles as sinks for radiation-induced vacancies and helium atoms is illustrated by a decrease swelling rate along the increase of the c_{He}/dpa ratio in Figure 5. The incubation period can be very well resolved in the plot (first two data points). Therefore, to describe analytically the role of helium production rate in the volumetric bubble swelling, we used the data plotted in Figure 5.

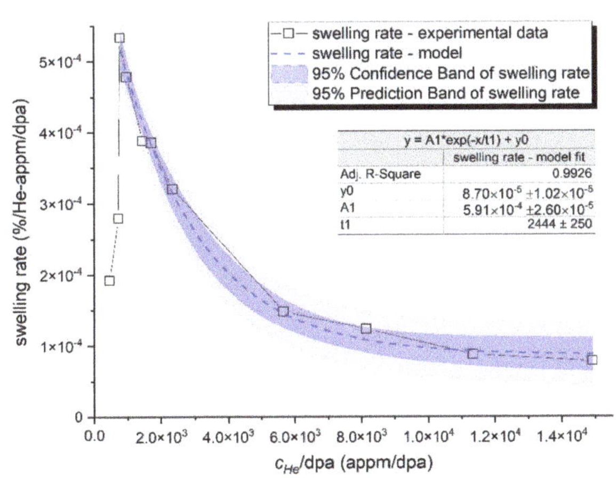

Figure 5. Swelling rate relative to the c_{He}/dpa ratio with a model fit to the data representing the volumetric bubble swelling in F/M steels exposed to radiation environments with high production rates of helium.

The following behavior, steady-state swelling and the saturation of the swelling, can be very well approximated by a swelling rate exponential decay formula $A * exp(-x/t_1) + y_0$, where A, t_1 and y_0 are approximately equal to 5.9×10^{-4}%/He-appm–/dpa, 2400 He-appm/dpa and 9×10^{-5}%/He-appm/dpa, respectively.

4. Conclusions

The present study provided a new perspective on the behaviour of reduced-activation ferittic/martensitic (RAFM) steels exposed to radiation environments having a high production rate of helium. A wide range of radiation conditions in helium production rates was investigated using a single-beam helium implantation and cross-sectional TEM analysis. Despite relatively low irradiation temperature (65 ± 5 °C) and a rather high helium concentration, the TEM showed a microstructure typical for RAFM steels exposed to harsh radiation environments. This suggested that low-temperature helium implantation can be a valuable tool in an experimental simulation of challenging radiation conditions, such as the environment of fusion reactors or spallation neutron sources. Our study suggested that a suitable depth-sensitive technique can provide unique data on helium-implanted structural materials to improve understanding of the role that the helium production rate plays on the evolution of the microstructure.

The results show virtually no swelling incubation period preceding a steady-state swelling of about 2×10^{-4}%/He-appm/dpa. A saturation of swelling above 5000 He-appm/dpa was observed and attributed to helium bubbles, which become dominant sinks for new vacancies and helium atoms. The results indicated that a high helium concentration introduced in near-room temperature implantations had a similar effect on the high irradiation temperature. The saturation of the volumetric swelling of helium-containing irradiated material, therefore, can be interpreted solely neither by displacement damage nor by the irradiation temperature, and all available information on helium concentration must be carefully considered.

Author Contributions: Conceptualization, V.K. and S.S.; validation, S.S., J.D. and V.S.; investigation, J.D., P.N., T.S. and S.S.; resources, P.N. and T.S.; writing—original draft preparation, V.K.; writing—review and editing, S.S. and V.K.; project administration, V.S.; funding acquisition, J.D. and V.K. All authors have read and agreed to the published version of the manuscript.

Funding: The authors would like to acknowledge partial support from the VEGA grant No. 1/0382/20 and VEGA 1/0395/20, as well as from the European Regional Development Fund, projects No. ITMS2014+: 313011U413 and 313011W085.

Institutional Review Board Statement: Non applicable.

Informed Consent Statement: Non applicable.

Data Availability Statement: All data used to reach the conclusions are presented in the paper. TEM micrographs are available upon request from the corresponding author.

Conflicts of Interest: The authors declare no conflict of interest.

References

1. Garner, F.A.; Toloczko, M.B.; Sencer, B.H. Comparison of swelling and irradiation creep behavior of fcc-austenitic and bcc-ferritic/martensitic alloys at high neutron exposure. *J. Nucl. Mater.* **2000**, *276*, 123–142. [CrossRef]
2. Dai, Y.; Krsjak, V.; Kuksenko, V.; Schäublin, R. Microstructural changes of ferritic/martensitic steels after irradiation in spallation target environments. *J. Nucl. Mater.* **2018**, *511*, 508–522. [CrossRef]
3. Zinkle, J.; Boutard, J.L.; Moelzer, D.T.; Kimura, A.; Lindau, R.; Odette, G.R.; Rieth, M.; Tan, L.; Tanigawa, H. Development of next generation tempered and ODS reduced activation ferritic/martensitic steels for fusion energy applications. *Nucl. Fusion* **2017**, *57*, 92005. [CrossRef]
4. Wang, X.; Yan, Q.; Was, G.; Wang, L. Void swelling in ferritic-martensitic steels under high dose ion irradiation: Exploring possible contributions to swelling resistance. *Scr. Mater.* **2016**, *112*, 9–14. [CrossRef]
5. Was, G. *Fundamentals of Radiation Materials Science: Metals and Alloys*; Springer: Berlin/Heidelberg, Germany, 2007; ISBN 978-3-540-49472-0.

6. Farrell, K.; Maziasz, P.J.; Lee, E.H.; Mansur, L.K. Modification of radiation damage microstructure by helium. *Radiat. Eff.* **1983**, *78*, 277–295. [CrossRef]
7. Noga, P.; Dobrovodsky, J.; Vana, D.; Beno, M.; Zavacka, A.; Muska, M.; Halgas, R.; Minarik, S.; Riedlmajer, R. A new ion-beam laboratory for materials research at the Slovak University of Technology. *Nucl. Instrum. Meth. B* **2017**, *409*, 264–267. [CrossRef]
8. Norgett, M.J.; Robinson, M.T.; Torrens, I.M. A proposed method of calculating displacement dose rates. *Nucl. Eng. Des.* **1975**, *33*, 50–54. [CrossRef]
9. Mansur, L.K.; Rowcliffe, A.F.; Nanstad, R.K.; Zinkle, S.J.; Corwin, W.R.; Stoller, R.E. Materials needs for fusion, Generation IV fission reactors and spallation neutron sources—Similarities and differences. *J. Nucl. Mater.* **2004**, *329–333*, 166–172. [CrossRef]
10. Kolluri, M.; Edmondson, P.D.; Luzginova, N.V.; Berg, F.A. A structure–property correlation study of neutron irradiation induced damage in EU batch of ODS Eurofer97 steel. *Mater. Sci. Eng. A* **2014**, *597*, 111–116. [CrossRef]
11. Klimenkov, M.; Lindau, R.; Materna-Morris, E.; Moslang, A. TEM characterization of precipitates in EUROFER 97. *Prog. Nucl. Energy* **2012**, *57*, 8–13. [CrossRef]
12. Klimenkov, M.; Moslang, A.; Materna-Morris, E. Helium influence on the microstructure and swelling of 9%Cr ferritic steel after neutron irradiation to 16.3 dpa. *J. Nucl. Mater.* **2014**, *453*, 54–59. [CrossRef]
13. Wang, K. Microstructure and fracture behavior of F82H steel under different irradiation and tensile test conditions. *J. Nucl. Mater.* **2016**, *468*, 246–254. [CrossRef]
14. Klueh, R.L. Analysis of swelling behaviour of ferritic/martensitic steels. *Philos. Mag.* **2018**, *98*, 2618–2636. [CrossRef]

materials

MDPI

Article

Positron Annihilation Study of RPV Steels Radiation Loaded by Hydrogen Ion Implantation

Vladimir Slugen [1,*], Tomas Brodziansky [1], Jana Simeg Veternikova [1], Stanislav Sojak [1,2], Martin Petriska [1], Robert Hinca [1] and Gabriel Farkas [1]

[1] Institute of Nuclear and Physical Engineering, Faculty of Electrical Engineering and Information Technology, Slovak University of Technology, Ilkovicova 3, 812 19 Bratislava, Slovakia

[2] Advanced Technologies Research Institute, Faculty of Materials Science and Technology, Slovak University of Technology, Jana Bottu 25, 917 24 Trnava, Slovakia

* Correspondence: vladimir.slugen@stuba.sk

Abstract: Specimens of 15Kh2MFAA steel used for reactor pressure vessels V-213 (VVER-440 reactor) were studied by positron annihilation techniques in terms of their radiation resistance and structural recovery after thermal treatment. The radiation load was simulated by experimental implantation of 500 keV H^+ ions. The maximum radiation damage of 1 DPA was obtained across a region of 3 μm. Radiation-induced defects were investigated by coincidence Doppler broadening spectroscopy and positron lifetime spectroscopy using a conventional positron source as well as a slow positron beam. All techniques registered an accumulation of small open-volume defects (mostly mono- and di-vacancies) due to the irradiation, with an increase of the defect volume $\Delta V_D \approx 2.88 \times 10^{-8}$ cm^{-3}. Finally, the irradiated specimens were gradually annealed at temperatures from 200 to 550 °C and analyzed in detail. The best defect recovery was found at a temperature between 450 and 475 °C, but the final defect concentration of about ΔC_D = 0.34 ppm was still higher than in the as-received specimens.

Keywords: VVER reactor; long-term operation; radiation-induced defects; recovery annealing; positron annihilation techniques

Citation: Slugen, V.; Brodziansky, T.; Simeg Veternikova, J.; Sojak, S.; Petriska, M.; Hinca, R.; Farkas, G. Positron Annihilation Study of RPV Steels Radiation Loaded by Hydrogen Ion Implantation. *Materials* **2022**, *15*, 7091. https://doi.org/10.3390/ma15207091

Academic Editor: Francisca G. Caballero

Received: 8 September 2022
Accepted: 3 October 2022
Published: 12 October 2022

1. Introduction

Prolongation of nuclear power plants' (NPPs) lifetime, also known as long-term operation (LTO), allows NPPs to generate reliable, low-cost, low-emission electricity for many years longer than originally envisioned and thus maximize their value. Today's operating nuclear reactors were originally designed for 30 to 40 years of operation, but there is no fixed technical limit to the life of the reactors [1]. LTO of NPPs has been successfully demonstrated and is increasingly recognized internationally as standard practice. LTO of an NPP may be conditioned by life-limiting processes and features of structures, systems, and components—the emphasis is on the irreplaceable reactor pressure vessel (RPV) and its construction steel. Degradation of its mechanical properties is the crucial limiting factor of NPP lifespan [2]. Elements causing degradation of RPV properties are neutron irradiation, high temperature and pressure, fatigue, corrosion, etc. Therefore, it is recommended to evaluate periodically the structural integrity of an RPV and to predict the future development of RPV degradation. It is well-known that neutron irradiation is the main driver of the microstructural changes in the RPV during reactor operation. RPV steels that are exposed to a wide spectrum of irradiation (including fast neutrons, whose influence is dominant) during LTO will degrade over time via effects connected to radiation damage. The traditional method of studying microstructural changes in reactor steels is based on stay (irradiation) in material research reactors, where (depending on position) irradiation of special model specimens is accelerated. Of course, although accelerated, these experiments take several months or years and cannot be equal in complexity to real conditions in commercial NPPs. Additionally, it is necessary to note that these experiments are not cheap and are difficult from a handling and radiation protection point of view. A promising approach could be the application of light ion accelerators for experimental simulation of neutron irradiation [3] performed on original commercial RPV steels.

In this paper, we will be focused mainly on positron annihilation techniques, which have been frequently applied in nuclear material microstructural studies in the last 50 years. Detailed descriptions of these techniques have been published worldwide in numerous scientific publications [4–7]. In the last 30 years, our laboratory has focused on the application of positron annihilation spectroscopy (PAS) techniques to nuclear reactor pressure vessel steels. In particular, RPV steels of Russian water-moderated and water-cooled pressured reactors (VVERs) were at the centrum of our studies. Although during the years we have come to understand the limits of these techniques, the conclusions and results interpretations we have made have resulted from combination with results from other techniques, and we continue with our trials to upgrade the level of experimental resolution as well as minimize the influence of possible disturbing factors as much as possible. An application of proton implantation as an experimental simulation of neutron irradiation seems appropriate in this study.

Nevertheless, we have touched also on point of limited reproducibility of results due to material inhomogeneities and a wide range of different possible errors.

Degradation of Reactor Pressure Vessel Steels

At the end of 2020, worldwide, 292 reactors have been in operation for longer than 30 years. This represents more than 65% of power reactors. During the last decade, only 63 new units were put in operation. From a long-term perspective and regarding the projected design lifetime, in 2030 about 160 reactors will be shut down [8,9]. Refurbishments and LTO seem to be a clever way to extend NPPs' lifetime and minimize global warming's impact on the world. It is important to stress that the operation of NPPs can deliver baseload power and can give a guarantee of quite acceptable and stable electricity costs.

Several NPPs, most notably 73 units in the United States, prolonged their operational license for up to 60 years. A similar trend is also present in other countries, where the crucial condition for operational lifetime prolongation is a decision from periodic safety reviews (PSRs). Based on these PSRs, which are supervised by a national nuclear regulatory authority, factors such as (i) aging management, (ii) environmental management, (iii) operating experiences, (iv) safety and security improvements, and emerging issues are considered and analyzed in detail [10].

The essential mechanisms of RPV irradiation embrittlement were deeply studied and analyzed via different RPV surveillance specimen programs. The RPV steel specimens were placed in specially designed capsules and irradiated in an operating reactor at exact analyzed conditions for several years. The base for evaluations was tensile, Charpy-V, and fracture toughness tests. After testing, these specimens were cut and used for additional non-destructive testing via different spectroscopic methods, and thanks to placement closer to the reactor core, an accelerating factor of up to 10 could be achieved [11].

Generally, it was stated that the most important embrittlement mechanisms are direct matrix damage, formation, and movement of Cu-rich precipitates (reported as CRP, with connection to Ni, Mn, Si, etc.), MnNi-rich precipitates (MNP), and P segregation on grain boundaries [12,13]. Recently published papers [14,15] also described Cu- and Ni/Si/Mn-rich clusters in irradiated reactor steels and they are often being located near line dislocations. Structural defects such as vacancies or dislocations probably can suppress the formation of large Ni/Si/Mn-rich and Cu-rich clusters by reducing the diffusion process under irradiation.

The neutron flux during the operation of commercial reactors depends on their design as well as power. Generally, in pressurized water reactors (PWRs), the neutron flux ranges between 10^9 and 10^{12} $n.cm^{-2}.s^{-1}$. The real impact of neutrons on irradiation damage in this relatively wide range of fluxes is influenced also by its spectrum, where special importance is placed on the fast neutrons (basically E > 1 MeV, in the case of VVERs neutrons with E > 0.5 MeV being considered) and differences in total neutron fluences. Of course, differences in coolant temperature (in the case of PWRs 260 to 330 °C), water chemistry, and reactor operation dynamics (out of all thermal pressurized shocks) can play a special role in real design material embrittlement. With an increase in radiation exposure, an increase in the number of obstacles can be observed too, and higher stresses are required for dislocation motion, with a resulting increase in the yield strength of the material. We have to consider also non-hardening embrittlement, which can appear via radiation-induced solute segregation to grain boundaries. This effect is usually assigned to the segregation of phosphorous on the grain boundary [16]. Many studies performed in the last 50 years have confirmed that the dominant effect on embrittlement from alloying elements involves copper. Its content dramatically influences irradiation sensitivity, more so than nickel, phosphorus, and vanadium, which are considered important contributors as well [11,16]. In addition to these elements, sulfur content, due to its negative influence on RPV steels' resilience, should be strictly controlled.

The duration of operational RPV lifetime is limited by neutron embrittlement. The crucial effect is resistance against potential brittle damage. Prolongation of NPP operation from the previously scheduled 40 years to the considered 60–80 years implies an increasing neutron exposure of about 100%. This lifetime extension could increase the maximal level of neutron fluences up to 1×10^{20} n.cm^{-2} [17]. The neutron irradiation shifts the Charpy toughness curve to increased levels of temperature. The second characteristic of the effect of increased irradiation is a significant decrease in fracture resistance [18].

One effective way to recover (at least partially) fracture toughness properties decreased due to irradiation of RPV steels is thermal annealing. Although we use the term annealing, it is correct to note that temperatures lower than 500 °C are not enough for complex annealing. The restoration of material toughness via thermal annealing was the focus of nuclear utilities mostly in connection to the VVER-440 V-230 design due to the not-so-strictly limited content of Cu and P in 15Kh2MFA steels, and annealing treatment was an acceptable solution for the increase of the operational safety margin as well as the pressurized thermal shock criteria requested by national regulatory authorities [19,20].

It is necessary to note that there exists a substantial difference between the old generation of the VVER-440 type V-230 and the second generation, called the V-213, regarding the chemical composition of steels. Therefore, the second generation, with minimized content of Cu and P, is marked as 15Kh2MFAA [21]. The differences are shown in Table 1, and an illustrative cross-section of typical VVER-440 RPV steel is depicted in Figure 1.

Table 1. Chemical composition (in wt.%) of 15Kh2MFA and 15Kh2MFAA RPV steels.

Steel	C	Mn	Si	P	S	Cr	Ni	Mo	V	Cu
15Kh2MFA	0.13	0.30	-	max	max	2.50	max	0.60	0.25	max
	0.18	0.60	-	0.020	0.025	3.00	0.40	0.80	0.35	0.30
15Kh2MFAA	0.11	0.30	0.17	max	max	2.00	max	0.60	0.25	max
Investigated steel	0.16	0.60	0.37	0.012	0.015	2.50	0.40	0.80	0.35	0.10
	Additions:	**As**	**Sb**	**Sn**	**P+Sb+Sn**	**Co**	**P**	**S**	**Cu**	
		0.10	0.005	0.005	0.015	0.02	0.012	0.15	0.08	

Based on results from several materials, studies of RPV steel specimens focused on macroscopical as well as microscopical changes due to annealing; the crucial conclusion was made from the tensile notch and Charpy-V tests. In the case of wet annealing, the knowledge was summarized in [22,23]. Changes in microstructure were observed mostly via transmission electron microscopy with relatively low resolution (corresponding to the time about 1980) at temperatures of ~340 °C minimum. This low temperature is an essential obstacle preventing significant minimization of irradiation embrittlement. Based on [23], the recovering effect is less than 50%. On the other hand, for so-called dry annealing with a much higher difference (≈230 °C) from the operating temperature at which RPV steels were long-term irradiated, the effectiveness level is over 90% [24,25]. The recent electron microscopy investigation of RPV steels and Fe-Cr model alloys after irradiation and annealing are described in more detail in [26–28].

Figure 1. Illustrative cross-section from VVER-440 RPV (Original not-irradiated peace from VVER-440 V-213 RPV—Greifswald Unit 7).

Especially important approaches for evaluation of RPV steels' long-term operational degradation were surveillance specimen programs [29,30], which provided the possibility to observe continuously the shift of ductile-brittle transmission temperature (DBTT) curves. Additionally, broken Charpy-V specimens were suitable for the preparation of new specimens for non-destructive testing via different methods of studying the phenomena but could contribute to the complex information about neutron embrittlement as well as the effectiveness of thermal annealing of RPV base and weld materials. In the last decade of the previous century, several spectroscopic methods (including positron annihilation techniques, whose applications will be reported in the next chapters) were used in these studies with the aim of describing microstructural changes, defect creation/relaxation in annealing, the role of alloying elements or long-term thermal and radiation treatment. From the positron annihilation sensitivity on vacancy-type defects point of view, irradiation damage studies were dominant.

From the complexity of irradiation that impacts RPV steels during decades of operation, we select neutron irradiation. Of course, there are also beta and gamma particles of different spectra or energies, but in comparison to the high-energy neutrons, we consider it plausible to neglect them. Similarly, alpha particles or potential light ions play no role due to distance from the core. Therefore, neutron irradiation is dominant in all considerations. The limiting factor of most conventional studies and methods is the activation of the specimens. Handling (cutting, surface cleaning) and possible contamination require special devices including hot cells, which make these studies expensive, time-consuming, and difficult from the radiation protection point of view. Experiments in material test reactors could not always achieve the requested conditions, mostly in neutron fluxes, fluences, temperatures, or irradiation dynamics.

Computer simulations were and still are frequently used in the last few decades but cannot replace material studies in an adequate form. Results from these simulations are sometimes unique, but at least some of them should be experimentally verified. There are fewer research reactors and, due to the limited possibility for experiments on power reactors, also less chance for verification. Therefore, experimental simulation via ion implantation (ion irradiation) seems to be one of the proper methods to conduct this type of study in the future. We have already started bearing in mind that verifying the equivalency of neutron and ion irradiation is a very important task [31–33]. For experimental neutron treatment, we have used proton (H^+) implantations due to our 0.5 MeV cascade accelerator.

There is no doubt that the degree of RPV embrittlement in a reactor pressure vessel (RPV) is a complex function of different parameters such as temperature, neutron fluence, flux, material chemistry, etc. [34–40].

2. Methodology

From the available positron annihilation techniques, we applied positron annihilation lifetime spectroscopy (PALS), coincidence Doppler broadening spectroscopy (CDBS), and a pulsed low-energy positron system (PLEPS).

Generally, the lifetimes of positrons, which are trapped in monovacancies, determined both experimentally and theoretically lie in the interval of 150–300 ps for metals [41]. The lifetimes of positrons trapped in small vacancy clusters have been calculated and experimentally verified many times. As the lattice relaxation around the cluster has a relatively small influence on positron annihilation parameters, it was not included in the calculations. The results show that the lifetime of a positron trapped by di-vacancy does not differ much from that for monovacancy (an increase is about 10 ps for Fe). The lifetime increases rapidly when the cluster grows into two-dimensional tri-vacancy and further into three-dimensional tetra-vacancy [41]. For very large clusters (over 50 vacancies), the lifetime saturates around 450–500 ps. For large voids, the trapping may be limited mainly by positron diffusion to the defect.

Theoretical calculations have revealed that the dislocation line is only a shallow trap for positrons [42,43] (binding energy <0.1 eV). On the other hand, the lifetimes of trapped positrons observed in plastically deformed metals [44] are only slightly lower than the lifetimes of positrons trapped in vacancies. Smedskjaer [45] suggested that the pure dislocation line is a weak positron trap and explained the long lifetimes seen in experiments by point-like defects (vacancies, jogs) associated with the dislocation. For example, the binding energy of 0.92 eV for a vacancy to the edge-dislocation line was calculated in [46] for Fe. Once a positron arrives at the core of the dislocation, it diffuses very quickly (pipe diffusion) until it finds a vacancy attached to the dislocation or a jog of the dislocation; it is then trapped and annihilates there. This explanation is supported also by further calculations [43,47–49]. Calculated lifetimes of positrons trapped in dislocations and corresponding binding energies are listed in Table 2.

Table 2. Calculated lifetimes τ and binding energies EB of positrons trapped in the core region of dislocation line and defects associated with dislocation [43,47,48].

Annihilation Site	τ	EB
	(ps)	(eV)
Fe: defect-free structure	97–110	
mono-vacancy	179	−2.39
di-vacancy	195	
vacancy with hydrogen	146	
Fe: edge dislocation line	117	−0.28
jog on the edge dislocation line	117	−0.11
vacancy on the edge dislocation line	140	−0.92
di-vacancy on the edge dislocation line	117	−0.07
Fe: screw dislocation	114	−0.22
vacancy in screw dislocation	174	−2.35

The specific trapping rates ν_D for dislocations were obtained by a combination of positron lifetime measurement and transmission electron microscopy (TEM) or other techniques capable of determining dislocation density (e.g., X-ray diffraction profile) [49]. The values of ν_D lie in the range of about 10^{-5}–10^{-4} m^2s^{-1} for metals. Thermally activated de-trapping of positrons trapped in a dislocation core (initial shallow trap) may occur at elevated temperatures, which makes ν_D temperature-dependent [50].

Table 2 shows calculated lifetimes τ (ps) of positrons trapped in bulk, vacancies, and dislocations. The different lifetimes for screw and edge dislocations in Fe single crystals were reported in detail in [48,49]. According to this work, screw dislocations exhibit larger specific trapping rates ν_D and lifetimes than edge ones. Similarly, to vacancies in the previous section, it can be shown that the minimum dislocation density detectable by PAS lifetime spectroscopy is $\rho_D \sim 10^{12}$ m^{-2}. On the other hand, if $\rho_D \sim 10^{16}$ m^{-2}, almost all positrons are trapped at dislocations (saturated trapping), and only the contribution of the trapped positrons is resolved in the PAS spectrum.

Positrons may be trapped in metals also by a grain boundary (GB). Nevertheless, trapping in GBs is likely only when the mean linear dimension of grains does not exceed a few µm. This means the grain size is comparable to (or smaller than) the positron diffusion length L+, and some fraction of positrons have a chance to reach a GB by diffusion motion. The transport of a positron to GBs limits substantially the positron trapping rate in GBs [51].

The relative values of the energy value for the ground state of a delocalized positron in different materials are different. This makes for possible positron trapping in a precipitate with a lower level of positron ground state energy. An excellent review paper about positron applications in radiation-induced defect studies has been published [52]. During the last few decades, many additional destructive and non-destructive methods have been applied with the aim of making progress in the characterization of complex information about defects' formation and their annealing and/or reannealing [53]. It is necessary to mention that some effects (neutron embrittlement and annealing of defects) are occurring at the same time, and the final microstructure depends on the level of neutron flux/fluence and the temperature of the treatment. Based on previous studies combining Mössbauer spectroscopy (MS) [54,55], positron annihilation spectroscopies (PAS) [56–70], and transmission electron microscopy (TEM) [62,64,68], we recognized that the PALS, as well as CDBS measurements, can yield substantial information about types and concentration of vacancies and their behavior after annealing [63–75]. Positrons trapped in open volume defects, i.e., radiation-induced vacancies, dislocations, microvoids, etc., annihilate with a lower probability than in the perfect area of the same material; thus, the positron lifetime grows with the size of the defects. The production of vacancy defects is directly attributed to radiation damage [59,63]. This was confirmed also by precise TEM studies [76,77]. Nevertheless, the interpretation of results depends on several uncertainties, and the reproducibility could be limited. Generally, we assume that specimens are homogenous. This condition, which is not fully fulfilled in the case of commercial RPV steels, is important for the application of the standard trapping model (STM) [78] in the evaluation of PALS measurements. In inhomogeneous specimens, positrons can be partially attracted by other trapping centers and the various implantation sites, which affects the positron data [79,80]. Therefore, the diffusion-trapping model (DTM) [81] was developed and successfully applied [82,83]. The DTM was several times improved, and its application for the pulsed positron beam technique is described in detail in [84]. The pulsed low-energy positron system (PLEPS) [85–87], which enables the measurement of irradiated specimens if their activities are lower than 1 MBq, was applied also for radiation degradation of RPV steels, although the pick/background ratio became significantly worse. This technique enables the depth profiling study of specimens (in the case of steels up to 550 nm) and can reduce the ^{60}Co radiation contribution to the lifetime spectra to a minimum.

3. Investigated Specimens, Experimental Treatment, and Experimental Techniques

In this work, Russian RPV steel specimens from 15Kh2MFAA (from Greifswald Unit 7, producers: Atomstroyexport JSC, Moscow, Russia & Škoda Works, Plzen, Czech Republic) commercially used in VVER-440 V-213 type reactors including the Mochovce 34 NPP, Slovenské elektrárne in Slovakia (today under physical start-up) are studied from the perspective of radiation damage and structural recovery after the annealing [88]. The investigated material was cut into 10 pieces, creating five pairs of specimens with dimensions of 10 mm × 10 mm × 0.2 mm. The specimens were polished in a mirror-like way to create perfect surface conditions before the surface radiation load. The chemical composition of the investigated specimens of 15Kh2MFAA steel was shown above in Table 1.

The radiation damage was experimentally simulated by proton irradiation (H^+) in a 500 kV implanter of the Research Centre for Ion Beams and Plasma Technologies SlovakION within the Slovak University of Technology in University Science Park CAMBO Trnava (Slovakia) [89]. Each specimen was implanted (irradiated) with protons with an energy of 500 keV and a hydrogen fluence of 1×10^{18} cm^{-2}. The total radiation damage achieved a value of up to 1 DPA. The layers were implanted at up to 3 μm according to the SRIM calculation for Fe-2.5Cr-0.6Mo implantation by H ions, and the highest level of radiation damage was at a depth of ~2.3 μm (Figure 2).

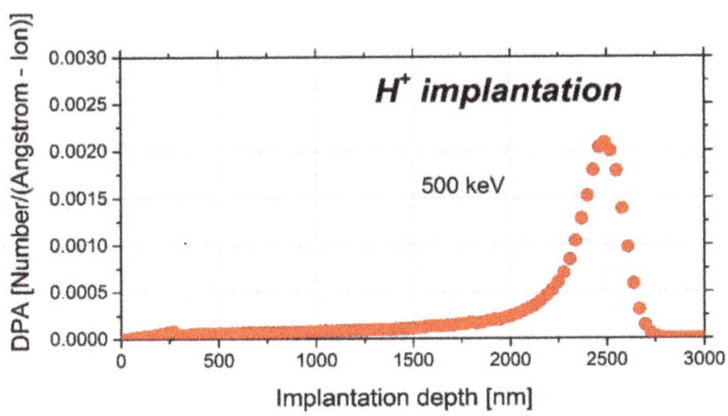

Figure 2. SRIM calculations of DPA for 500 keV H^+ ions incident on Fe-2.5Cr-0.6Mo. The detailed K-P calculation was used with a total number of ions in the calculation of 5×10^4.

The investigated specimens were observed by positron annihilation techniques due to their strong sensitivity to small vacancy defects formed during the process of irradiation/implantation. The radiation load affected mostly surface and subsurface layers, and therefore slow positrons were firstly applied for verification of a defect accumulation and observation of a defect depth profile up of to 520 nm. For that, a pulsed low-energy positron system (PLEPS) with the high-intensity positron source NEPOMUC at the FRM-II reactor in Technical University of Munich, Garching, Germany [90,91] was used.

Later, positron annihilation lifetime spectroscopy (PALS) and coincidence Doppler broadening spectroscopy (CDBS) [92] with a conventional positron source of ^{22}Na (Institute of Nuclear and Physical Engineering, Slovak University of Technology, Bratislava, Slovakia) were applied at the Institute of Nuclear and Physical Engineering, Bratislava. The measurements were performed with expectations that only ~20% of positrons can provide a measurable signal from the implanted region up to 3 μm [93], while these techniques investigate materials up to 150 μm.

A PALS spectrum with a total count of $\approx 10^6$ was measured by three Hamamatsu H3378 photomultipliers tubes coupled with BaF2 scintillators (Hamamatsu Photonics, Hamamatsu City, Shizuoka Pref., Japan) and powered by Ortec 556 HV sources (ORTEC/AMETEK, Oak Ridge, Tennessee, USA) in an air-conditioned box (Figure 3a) at room temperature. The START signal for the lifetime measurement is ~1.2 MeV gamma photons emitted by the positron source ^{22}Na. The STOP signal, the annihilation energy of 511 keV, terminates the timing of the positron lifetime measurement, which is directly recorded and stored by DRS4 digitizer chip (RADEC, Koblenz, Switzerland). The own software QtPALS (Petriska, M., Institute of Nuclear and Physical Engineering, Slovak University of Technology, Bratislava, Slovakia) was used for pulse processing and spectrum construction. The data were treated by LT 9 software (Kansy, J., Institute of Physics and Chemistry of Metals, Katowice, Poland).

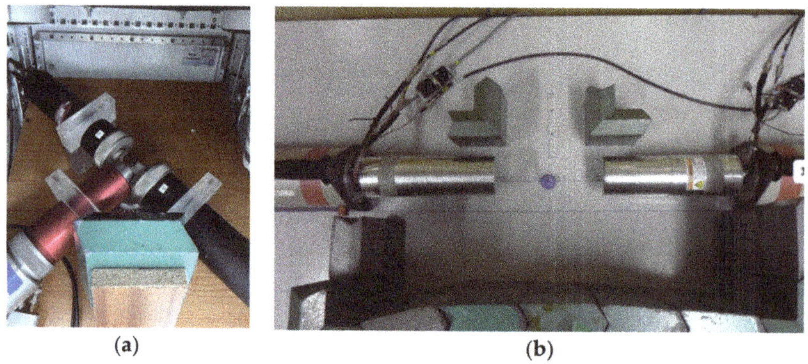

Figure 3. Air-conditioned unit setup for PALS (**a**) and CDB instrument in Institute of Nuclear and Physical Engineering, Bratislava (**b**).

CDB techniques [94] observed spectra with about 10^6 counts by two HPGe detectors GC2019 (ORTEC/AMETEK, Oak Ridge, Tennessee, USA) with Gaussian resolution function 1.9 keV at 1.33MeV (Figure 3b) cooled by the cooling system Cryo-JT. As a high voltage source, a dual high-voltage power supply Canberra3125 (Canberra Industries, Inc./Mirion Technologies, Inc., Meriden, CT, USA) is used. The standard multi-channel analyzer and spectra amplifier was replaced by digital components—DAQ Adlink-PCI (ADLINK Technology, Inc., Taoyuan City, Taiwan). The momentum window for a calculation of the S parameter is $|pL| < 2.5 \times 10^{-3} m_0c$, and for the W parameter, it is $15 \times 10^{-3}\ m_0c < |pL| < 25 \times 10^{-3}\ m_0c$. Pulse amplitudes were collected and evaluated by own software – QtPALS and QtCDB (Petriska, M., Institute of Nuclear and Physical Engineering, Slovak University of Technology, Bratislava, Slovakia).

After the observation of the radiation damage, the specimens were gradually annealed at 200, 300, 400, 450, 475, 500, and 525 °C for detection of the optimal annealing temperature in the process of structural recovery. The post-irradiation annealing was performed in a compact vacuum tube furnace MTI GSL-1800x (MTI Corporation, Richmond, CA, USA). The specimens were annealed for 2 h in a ~6.4×10^{-9} vacuum bar and after that cooled down in the air for about 45 min. The specimens were then observed by the PALS technique, whereas the annealing process affected the whole depth of the specimen.

4. Results

4.1. PLEPS Measurement

PLEPS measurement of one investigated specimen was used to confirm the ability of protons to deliver significant lattice damage and observation of the defect depth profile up to 520 nm. The specimen in the as-received state and after the implantation were compared in terms of defect size and defect presence in the surface and subsurface layers.

Figure 4 shows that positron lifetimes in bulk (LT_1) of the as-received specimen remained approximately at the same level through all measured depths, which is in perfect agreement with expectations because positron annihilation in bulk should remain the same independently of the state of the material. However, LT_1 for proton-irradiated specimens compared to the non-irradiated ones increased from 140 ps to 160–170 ps. A slight increase of the positron lifetime LT_1 in our measurements compared to positron lifetime LT_1 in pure Fe (~107 ps [95]) might be attributed to the fact that these spectra were also decomposed into two components instead of the usual three components; thus, the DTM model was applied.

Because of the two components' decomposition, we believe the evaluation software attributed some annihilations to minor lattice defects such as dislocation and monovacancies to the annihilation in bulk, which resulted in the increase of LT_1. The same explanation applies also to the ~20–30 ps difference between the non-irradiated and the irradiated states—the presence of larger defects in lattices after proton irradiation led to their attribution to annihilation in bulk and therefore the increase of LT_1 compared to the as-received state.

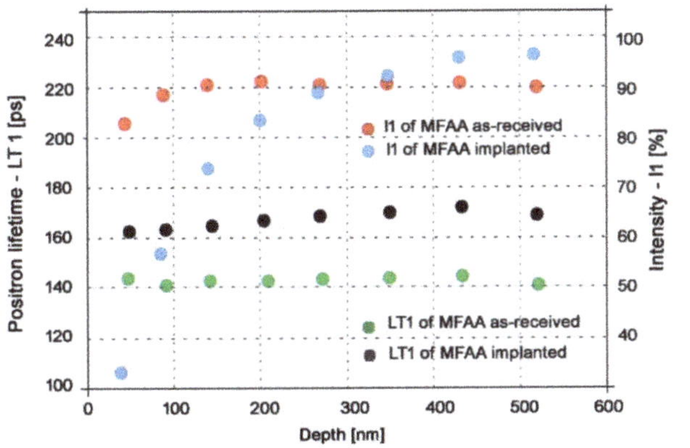

Figure 4. Positron lifetimes (LT_1) and intensity of annihilation (I_1) in bulk for various energies of incident positrons (penetration depth) for both non-irradiated and proton-irradiated specimens.

The intensity of annihilation in bulk (I_1) for the depth range of 150–520 nm for the as-received state remains constant at values around 90%; only for depths of 50 and 90 nm is I_1 is somewhat lower. This decrement is naturally accompanied by a proportional increase in the intensity of annihilation in defects (I_2), as seen in Figure 5. This can be attributed to the fact that positrons after implantation and thermalization move randomly, and during this process, they can also return to the entrance surface.

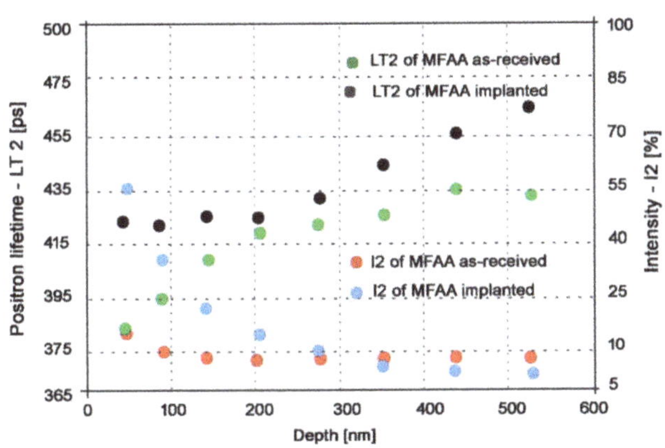

Figure 5. Positron lifetimes (LT_2) and intensity of annihilation (I_2) in defects for various energies of incident positrons (penetration depth) for both non-irradiated and proton-irradiated specimens.

"Positrons can be trapped at the surface where the electron density is lower than in the bulk. Additionally, they can be trapped at surface defects or positronium, i.e., a bound state between positron and electron can be created" [96]. Another factor that might be contributing is residue defects from specimen preparation—ones that remained even after polishing. All of this causes the I_1 value to be lower and the I_2 value to increase at depths closer to the surface than in the bulk (Figure 5). This effect is even more significant for proton-irradiated specimens. As seen in Figure 4, I_1 decreases from ~95% for depths deep in the bulk to 45% for depths close to the surface (also accompanied by a proportional increase in I_2 in Figure 5). We believe that this significant decrease in I_1/increase in I_2 for proton-irradiated specimens was caused mainly by storage of the specimens (creation of an oxidation layer on the surface, see below) and by manipulation and transportation of specimens during proton irradiation and PLEPS measurements (surface damage).

As seen from Figure 5, positron lifetimes in defects (LT_2) for depths of 50, 100, and 150 nm is 20–40 ps higher for the proton-irradiated specimens compared to the as-received state. At depths this close to the surface, the possibility that this disproportion would be created by proton irradiation is excluded because, as shown by SRIM simulations, the lattice damage starts at depths of ~100–150 nm (Figure 5). This increase in LT_2 was probably caused by the creation of an oxidation layer on the surface of the specimen. The assumption of the oxidation layer creation is supported by the results from other papers focused on the study of corrosion-related defects by slow positron beam. These papers reported similar values of LT_2 in the surface oxidation layer [97,98]. However, the creation of the surface oxidation layer was expected because the period between specimen preparation and PLEPS measurements of proton-irradiated specimens on this occasion was prolonged because of various problems with the FRM-II reactor to more than two years. Figure 5 also shows that the surface oxidation layer reaches a depth of around 200 nm, where the positron lifetime in defects (LT_2) for proton-irradiated specimens exhibits the same values as for the non-irradiated state (without the oxidation layer).

For depths starting at ~270 nm, a progressive increase in LT_2 for the proton-irradiated state is clear compared to the non-irradiated state. This progressive increase in LT_2 is in perfect agreement with SRIM simulations because, as seen in Figure 2, the number of vacancies produced starts to increase at this depth. At maximum depth, $LT_{2,}$ reaches ~460 ps, which indicates the presence of vacancy clusters containing more than 40 vacancies [99]. It is important to note that the maximum depth of positrons reached by PLEPS is, as seen in Figures 4 and 5, around 0.52 μm, but according to SRIM simulations, the maximum lattice damage is at 2.3 μm; therefore, even bigger vacancy clusters are expected to form deeper in the material. Unfortunately, we were unable to probe deeper into the specimen because 0.52 μm is the maximum depth that can be reached by positrons in PLEPS. By using 500 keV protons, we hit the spot where this energy is too low for effective usage of conventional PALS and simultaneously too high for PLEPS to see the overall lattice damage. Nevertheless, PLEPS data confirmed that lattice damage was introduced to the material by proton irradiation.

4.2. CDB Measurement

The CDB measurements using standard positron energy spectra were preliminarily applied only for five as-received and five implanted specimens as confirmation of PLEPS results and the test of sensitivity to the observed defects by a conventional positron source. The CDB results characterizing the material at a depth of 150 μm are shown in the S-W diagram (Figure 6), where the W parameter describes positron annihilation with core electrons (annihilation in bulk) and the S parameter with valence electrons located mostly in open volume (annihilation in vacancy defects). These two parameters are extracted from each spectrum, and each carries different information about the measured materials [4].

As seen in Figure 6, proton-irradiated specimens exhibit a significant increase in the S parameter and a decrease in the W parameter compared to the non-irradiated specimens. The total amount of annihilating positrons is fixed, and in general, an S parameter increase is necessarily accompanied by a W parameter decrease. The S parameter can be affected by both the number density and the size of vacancy-type defects [100], so in our case, this enhancement of the S parameter indicates defect accumulation in the structure due to the implantation. These results are in good agreement with the PLEPS results above, and they also indicate that conventional positron sources can be useful for the detection of implantation-accumulated defects in our structure.

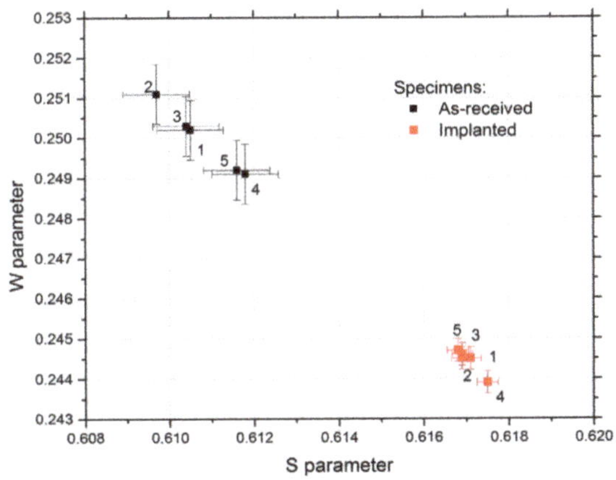

Figure 6. W parameter as a function of the S parameter for the as-received and proton-irradiated state.

It is important to note that the data's variance for all as-received specimens is 0.35%, and for all irradiated ones, it is 0.11%, i.e., the scattering of the measured data is very low. This proves that the results are not very sensitive to material inhomogeneity, clusters of carbides, disparities in specimen preparation, or inaccuracies during experimental measurements and during proton irradiation of the specimens. Therefore, the standard trapping model [78] can be applied for the evaluation of the PALS spectra.

The average S parameter of the as-received specimens is 0.6108 ± 0.0007, and the average W parameter is 0.24998 ± 0.0007. The average CDB parameters changed due to the implantation, and they are S = 0.617 ± 0.0003 and W = 0.2444 ± 0.0003. The implantation caused the S parameter to increase about 1.01 times, which indicates an accumulation of very small defects (mono- or di-vacancies).

4.3. PALS Measurement

All five investigated specimens were observed by the PALS technique in the as-received state, after the implantation, and after the gradual annealing. They showed good homogeneity in structure, and the response to the experimental loads was also similar, as can be seen in Figure 7. Specimen no. 5 is measured only up to the annealing temperature of 500 °C due to the specimen sustaining mechanical damage during the experiment.

PALS data were evaluated in the program LT9 by the standard trapping model [90] and separated into two or three components including lifetimes (LT) and intensities (I). For the first lifetime (LT_1), the bulk was fixed to 100 ps (calculated values for a pure iron range from 97 ps [95] to 110 ps [48]) for better comparison of the second components for all measurements. The second component involves small defects and can describe their size as proportional to a lifetime (LT_2, mostly dislocations and vacancies) and their concentration which is proportional to the intensity (I_2). The third component (LT_3, I_3) was found only for implanted specimens and annealed specimens at temperatures out of range (300–500) °C, where we expected the most visible change in structure. The LT_3 was on average 1.4 ± 0.5 nm with I_3 of approx. 1% and describes annihilation in the source or infight annihilation not fully compensated during the process of fitting.

The LT_2, I_2, and positron Mean Lifetime values (MLT, describing the whole spectra while minimizing the influence of the data treatment) are presented in Figure 7 for all measurements. For the individual treatments of specimens, the average of LT_2 and I_2 as well as their standard deviations were calculated considering the uniformity of the specimens and their good structural homogeneity. These average values are described below in more detail.

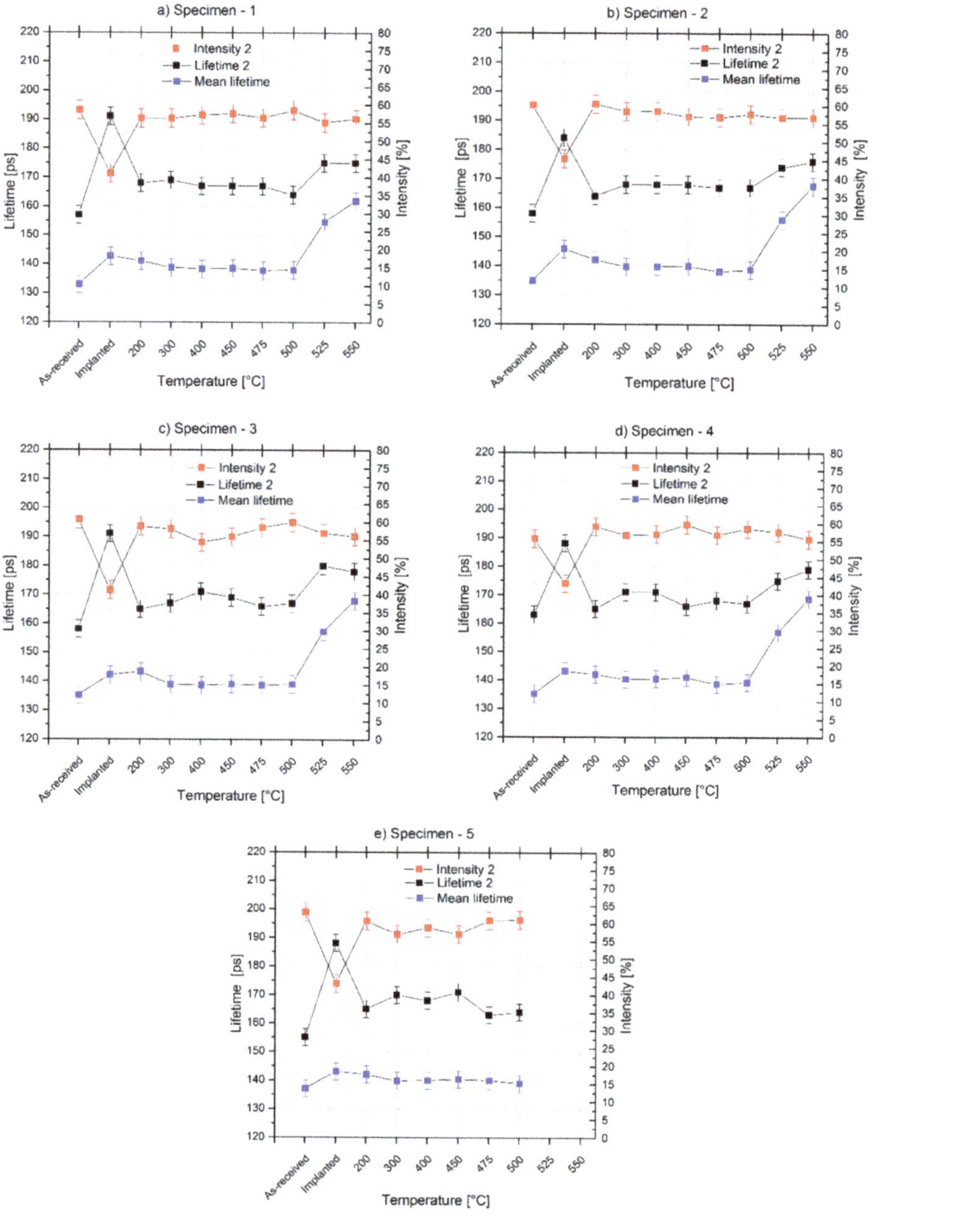

Figure 7. PALS data for as-received, implanted, and annealed specimens: Lifetime 2 proportional to the size of the defects; Intensity 2 proportional to defect concentration; and Mean lifetime describing the defect presence in the specimens.

In the as-received state, the average LT_2 = 158 ps ± 3 ps probably represents dislocations together with the smaller presence of mono-vacancies. The average intensity of as-received specimens was 59.7% ± 2.5%. The Mean Lifetime for the as-received specimens was between 134 and 137 ps, and the average value was 135 ps ± 2 ps. It is the lowest value for all our measurements, which indicates the lowest total defect volume (Figure 8), as was expected.

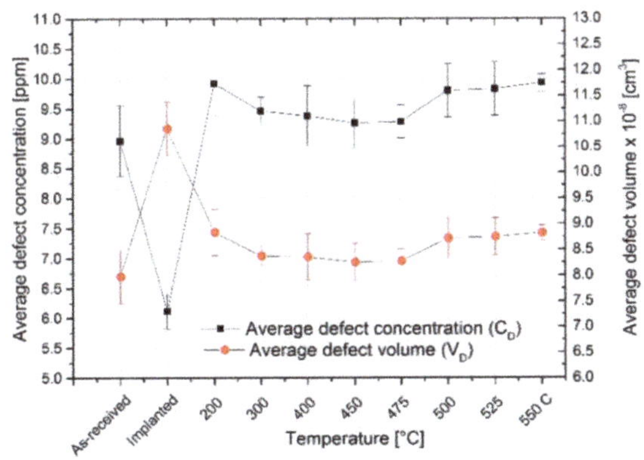

Figure 8. Average values of defect concentration and defect volume for the as-received, implanted, and annealed specimens at 200, 300, 400, 450, 475, 500, 525, and 550 °C.

After the implantation, LT_2 increased, with the average value of 188 ± 3 ps representing a mixture of mono-vacancies and di-vacancies with intensity (I_2) 42.8 ± 1.6%. The third lifetime appeared with average values of 1 544 ± 820 ps and intensity of 0.4 ± 0.1%. The MLT of the implanted specimen was between 142 and 146 ps, and the average value was 143 ± 2 ps. This is a much higher MLT value than for the as-received specimens. The MLT change was 8 ± 4 ps, which proves there was defect accumulation due to H^+ implantation in the specimen. The mono-vacancies partially changed into di-vacancies during the process of irradiation. Although previously found dislocations did not disappear from the structure, only the positrons were more attracted and trapped in the larger defects, di-vacancies, which could be near the dislocations [45].

The second components of PALS spectra and MLT were progressively changed after the gradual experimental annealing at 200, 300, 400, 450, 475, 500, 525, and 550 °C, as is seen in Figure 7. The annealing temperature of 200 °C did not show significant change from the perspective of MLT, 142 ± 1 ps, but there were found mostly mono-vacancies, probably together with dislocations again. The average value of the second component for the specimens annealed at 200 °C is LT_2 = 165 ± 3 ps and I_2 = 59.1 ± 1.5%. The existence of the third component affected the MLT value, and thus its change could probably not be so evident. The average values for third component are LT_3 = 992 ± 190 ps and I_3 = 0.4 ± 0.09%.

The specimens annealed at temperatures from 300 to 500 °C showed a small decrease in defect presence compared to the implanted data as well as the specimens annealed at 200 °C. The average MLT values for all annealed specimens are similar, being 140 ± 2 ps (Figure 7). Only the specimens annealed at temperatures between 475 and 500 °C had MLT values of 139 ± 1 ps, which is a practically negligible difference. However, the temperature of 475 °C is commercially used for the recovery of reactor steels in older VVER reactors. Our radiation-induced changes had much lower levels and were only present in the thin layer of the specimens, and thus the change here is not so significant.

The second components for specimens annealed at temperatures between 300 and 500 °C are almost the same and showed the presence of mostly mono-vacancies together with dislocations: LT_2 = 169 ± 2 ps and I_2 = 57.3 ± 0.9% (at 300 °C), LT_2 = 169 ± 2 ps and I_2 = 56.8 ± 1.4% (at 400 °C), LT_2 = 168 ± 2 and I_2 = 57.6 ± 1.3% (at 450 °C), LT_2 = 167 ± 2 ps and I_2 = 57.1 ± 1.8% (at 475 °C), LT_2 = 166 ± 2 ps and I_2 = 56.68 ± 0.9% (at 500 °C).

The MLT for specimens annealed at higher temperatures than 500 °C indicates an increase in the defect presence at a more significant rate than during the implantation. The second component for specimens annealed at 525 °C has average values LT_2 = 176 ± 3 ps (mono-vacancies) and I_2 = 57.3 ± 0.9%. The second component for specimens annealed at 550 °C is LT_2 = 177 ± 2 ps (mono-vacancies) and I_2 = 56.3 ± 0.6%. These specimens have also third components with average values: LT_3 = 1 228 ±150 ps and I_3 = 1.19 ± 0.2 (at 525 °C), LT_3 = 1 865 ± 180 ps and I_3 = 1.3 ± 0.2 (at 550 °C).

From the positron data, the defect concentrations and defect volume were calculated according to equations published in [101–103] and presented in Figure 8. The average values of defect concentration and defect volume for the individual treatments are seen in Figure 8. The calculation was applied only for the material visible by positrons up to a depth of approximately 150 μm, which is a volume of around 0.012 cm^3 with ≈ 10^{21} atoms.

The average defect concentration (C_D) decreased about ΔC_D = − (2.85 ± 0.89) ppm after the implantation due to the formation of larger defects (from dislocation with mono-vacancy to mostly di-vacancy). However, the average defect volume (V_D) observed by positrons increased by ΔV_D = + (2.88 ± 1.05) × 10^{-8} cm^3 due to the implantation.

After annealing at 200 °C, the size of defects decreased, and the defect concentration increased back by ΔC_D = + (3.84 ± 0.83) ppm compared to the implanted specimens and by ΔC_D = + (0.99 ± 0.73) ppm for to the as-received specimens. This indicates the presence of more defects than in the two previous stages of the specimens. The average defect volume decreased by ΔV_D = − (2.00 ± 0.97) × 10^{-8} cm^3 compared to the implanted specimens due to partial recombination of di-vacancies and a shift of the average defect size from di-vacancy to a vacancy. The change of the defect volume compared to the as-received specimens is ΔV_D = + (0.88 ± 0.72) × 10^{-8} cm^3, and thus the annealing did not recover the material.

The decline of defect volume due to annealing was the biggest at 200 °C, although the defect volume still gradually decreases up to 500 °C, almost only within the error bars. However, the absolute lowest values for defect concentration and defect volume within the annealed specimens seem to occur at temperatures 450 and 475 °C. The values ΔC_D ≈ − 3.19 ppm and ΔV_D ≈ − 2.57 × 10^{-8} cm^3 showed the biggest decrease compared to the implanted specimens for both temperatures (the same values). The changes in the values for 450 and 475 °C compared to the as-received specimens are ΔC_D ≈ + 0.34 ppm and ΔV_D ≈ + 0.30 × 10^{-8} cm^3, which shows that no full recovery of the structure occurred due to the annealing.

At the higher temperatures of 500, 525, and 550 °C, the defect concentration and the defect volume visibly increased more than ΔC_D ≈ 0.85 ppm and ΔV_D ≈ 0.75 × 10^{-8} cm^3 compared to the as-received specimens, but the defect volume is still lower than for the implanted specimens due to smaller defects. This could be due to thermal strain and the formation of new vacancies due to structural changes (precipitates, defect mobility, thermo-vacancies).

5. Discussion

Degradation of design materials of nuclear installations via aging or neutron embrittlement is a permanent and long-term studied nuclear-relevant problem. In addition to conventional methods (destructive tests), several non-destructive techniques including positron annihilation were involved in these studies. The reproducibility of PAS results after more than 30 years of application of these techniques worldwide (with the sometimes-disputable origin of specimens or only having a few specimens or based on laboratory-prepared binary alloys) was discussed and analyzed in this paper.

Due to difficulties related to the traditional neutron irradiation experiments (expense, time consumption, special equipment requirements, and radioactivity), the first research aimed to study proton irradiation of 15Kh2MFAA steel as a potential surrogate of neutron irradiation in experiments related to LTO. Proton irradiation provides a faster, cheaper, less dangerous, and repeatable way to investigate basic radiation damage processes taking place in RPVs. 15Kh2MFAA is the reactor pressure vessel steel that was used as the structural material for the Russian VVER-440/213 reactors. Specimens of 15Kh2MFAA steel obtained from the surveillance program of the Mochovce 34 NPP were irradiated by light ions (500 keV protons) to emulate radiation damage from neutrons. After irradiation, specimens were annealed at various temperatures to determine the level of lattice recovery at each temperature. Three different non-destructive positron annihilation-based techniques (positron annihilation lifetime spectroscopy—PALS, coincidence Doppler broadening spectroscopy – CDBS, and pulsed low-energy positron system—PLEPS) were used to investigate and characterize the microstructural properties of the 15Kh2MFAA steel. The characterization was based on the determination and comparison of the size and concentration of vacancy-type defects in the material.

In this study, 500 keV proton implantation delivered only slight lattice damage to the structure, and therefore the annealing experiment did not provide satisfactory results. However, we were able to detect slight differences in positron data for the as-received state, irradiated state, and annealed state at all investigated temperatures.

PLEPS and CDB results provided evidence for changes in microstructure due to the implantation. The PALS data indicate that the implantation process can be also observed by conventional methods, although the data changes are often very indistinct and should be confirmed or completed by another technique, such as PLEPS and CDB in this study. The increase of MLT due to the implantation was $\approx$8 ps, which represents a small difference, but enough for an indication of a change in structure and defect presence.

The annealing experiment observed by the PALS technique found smaller changes, but they progressed with the increase of the annealing temperature. While the temperature of 200 °C showed the biggest decrease in the defect volume after the implantation, the absolute lowest values of defect volume and concentration were found at the temperatures of 450 and 475 °C. In terms of the Mean Lifetime, annealing between 475 and 500 °C seems to be the best for structural recovery.

It is known from accumulated research that carbon atoms within the matrix play a key role in terms of vacancy cluster formation in Fe-C alloys and steels. Carbon and vacancies strongly interact and form stable carbon-vacancy (CV) complexes such as CV, C_2V, C_4V_2, and bigger ones. These CV complexes act as traps for radiation-induced vacancies, leading to the enhanced formation of vacancy clusters [104–107]. Therefore, the small recovery observed in the temperature interval of 200–300 °C could be explained by the decomposition of the small carbon-vacancy pairs or complexes. These findings correspond well with the available literature, where various authors note that the dissolution of vacancy complexes occurs at this temperature range in steels and Fe-C alloys [108–111]. However, this phenomenon of annealing CV complexes probably does not play a key role in terms of lattice recovery of VVER RPVs after irradiation because of the normal operating temperature (270–300 °C), and therefore it is expected that those types of defects will recombine shortly after their origin.

Furthermore, the PALS data for the annealing at 500, 525, and 550 °C proved that higher temperatures affect the structure via thermal structural changes. The increase in positron lifetime above 475 °C corresponds to the nucleation and growth of metal carbide/nitride precipitates, with a majority of the precipitates being MC, M_7C_3, and $M_{23}C_6$ carbides [96,112–114]. The carbide precipitation begins at 450 °C with vanadium-rich carbides (MC). Simultaneously, or at somewhat higher temperatures iron–chromium carbides (M_3C) precipitate. These are replaced by M_7C_3 carbides at higher temperatures [113]. With the increase in the number density and the size of precipitates, the associated precipitate–matrix interface area increases, and hence, the interfacial open volume defects increase. The increase in the lifetime could be also due to vacancy defects inside the precipitates [112]. The PALS results for higher temperatures confirmed that the ideal annealing temperature for 15Kh2MFAA is up to 475 °C. Similar behavior of positron lifetime in this region for RPV or model steels has been reported by various authors [112,115–117].

We assume that open volume defects as vacancies and interstitial points are mobile and recombine together at the normal operating temperatures of most RPVs. However, it is also expected that they can interact with solute atoms. The key interstitial impurity in Russian RPV steel is carbon [118,119]. The partial or complete trapping of self-interstitial points by C solutes will cause heterogeneous cluster nucleation and fine cluster distribution. Furthermore, in steels containing residual levels of elements such as copper, which are in super-saturated solution, radiation-enhanced diffusion will occur at these temperatures, which leads to the formation of small clusters, which can again harden the matrix. Generally, thermal treatment together with neutron irradiation led to a microstructure consisting of small clusters (<5 nm in diameter) which create obstacles to the free movement of dislocations, thereby producing an increase in the yield stress, hardness, and ductile-brittle transition temperature of the material.

It was shown that PLEPS can see the formation of solute clusters during irradiation of RPV steels, resulting in the depletion of Cu and P in the matrix in the first period of irradiation [86,95,118]. Some alloying elements (for example, Ni) can slightly retard this depletion. After this first period, both techniques registered no significant changes connected to increased neutron treatment. On the contrary, the positron lifetimes decreased, probably due to long-term thermal treatment at the level of 280 °C. Using PALS, the most effective region (450–475 °C) of thermal treatment was clearly shown.

Positron annihilation techniques can be applied for the development of new types of steel for advanced nuclear facilities such as fusion reactors or spallation neutron sources [120]. They are particularly useful for the evaluation of the effectiveness of post-irradiation thermal treatments [78]. The future is open to new investigative methods, and an application of a scanning positron microscope for testing RPV steel would surely be a good method [121].

6. Conclusions

The present paper reports a multi-specimen positron annihilation study of reactor pressure vessel steel 15Kh2MFAA in the as-received state, after proton implantation, as well as after isochronal annealing at temperatures (200–550) °C. Based on the results, the overall reproducibility of the obtained data was determined to be better than 2 ps in terms of the positron Mean Lifetime. The proton implantation, performed as an experimental simulation of neutron exposure, led to apparent differences in positron lifetime and Doppler broadening (of the annihilation gamma line) spectra. The presence of radiation-induced defects was observed via techniques utilizing slow positron beams as well as via techniques based on radioisotope positron sources.

According to our experimental results, it is possible to conclude that all used positron annihilation techniques registered the accumulation of small open-volume defects (mostly mono- and di-vacancies) due to the simulated irradiation (proton implantation), with the increase of the defect volume being $\Delta V_D = 2.88 \times 10^{-8}$ cm^{-3}.

The annealing experiments were performed on radiation-loaded specimens in the temperature range from 200 to 550 °C and analyzed. The best defect recovery was found at temperatures between 450 and 475 °C, but the defect concentration was still higher there, about $\Delta C_D = 0.34$ ppm higher than in the as-received specimens.

Author Contributions: Conceptualization, V.S. and T.B.; methodology V.S., R.H., G.F. and M.P.; validation, R.H. and G.F.; investigation, S.S., T.B. and M.P.; software, T.B., J.S.V., S.S. and M.P.; writing, V.S., T.B., J.S.V. and S.S.; writing (review and editing), S.S., M.P. and J.S.V.; visualization, J.S.V.; supervision V.S.; project administration, V.S.; funding acquisition, V.S. All authors have read and agreed to the published version of the manuscript.

Funding: This research was funded by the European Commission—Project DELISA-LTO (No.101061201).

Institutional Review Board Statement: Not applicable.

Informed Consent Statement: Not applicable.

Data Availability Statement: The data presented in this study are available on request from the corresponding author.

Conflicts of Interest: The authors declare no conflict of interest.

References

1. *The Enduring Value of Nuclear Energy Assets*; Report No. 2020/003; World Nuclear Association Publishing: London, UK, 2020; p. 12.
2. Kryukov, A.; Debarberis, L.; Ballesteros, A.; Krsjak, V.; Burcl, V.; Rogozhkin, S.V.; Nikitin, A.A.; Aleev, A.A.; Zaluzhnyi, A.G.; Grafutin, V.I.; et al. Integrated analysis of WWER-440 RPV weld re-embrittlement after annealing. *J. Nucl. Mater.* **2012**, *429*, 190–200. [CrossRef]
3. Slugeň, V.; Sojak, S.; Egger, W.; Krsjak, V.; Simeg Veternikova, J.; Petriska, M. Radiation Damage of Reactor Pressure Vessel Steels Studied by Positron Annihilation Spectroscopy - A Review. *Metals* **2020**, *10*, 1378. [CrossRef]
4. Krause-Rehberg, R.; Leipner, H.S. *Positrons Annihilation in Semiconductors – Defect Studies*; Springer: Berlin, Germany, 1997; p. 375. ISBN 3-540-64371-0.
5. Hautojarvi, P.; Dupasquier, A.; Manninen, M.J.; Vehanen, A.; Mijnarends, P.E.; West, R.N.; Nieminen, R.M. *Positrons in Solids*; Hautojarvi, P., Ed.; Springer: Berlin/Heidelberg, Germany, 1979; p. 255. ISBN 978-3-642-81318-4.
6. Hautojärvi, P.; Corbel, C. *Proceedings of the International School of Physics "Enrico Fermi", Course CXXV*; Dupasquier, A., Mills, A.P., Eds.; IOS Press: Varenna, Italy, 1995; p. 491.
7. Seeger, A.; Major, J.; Jaggy, F. *Positron Annihilation*; Jain, P.C., Singru, R.M., Gopinathan, K.P., Eds.; World Scientific Publishing: Singapore, 1985; p. 137.
8. The Database on Nuclear Power Reactors. PRIS–Power Reactor Information System, IAEA Website. Available online: https://pris.iaea.org/pris/home.aspx (accessed on 20 January 2022).
9. Nick, K.S.; Bourdon, P. *Legal Frameworks for Long-Term Operation of Nuclear Power Reactors*; Legal Adviser, OLC; OECD Publishing: Berlin, Germany, 2019; ISBN 978-92-64-63128-1.
10. Lokhov, A.; Huerta, A.; Dufresne, L.; Giraud, A.; Osouf, N. (Eds.) *The Economics of Long-Term Operation of Nuclear Power Plants*; OECD/NEA Publishing: Berlin, Germany, 2012; p. 116. ISBN 978-92-64-99205-4.
11. Altstadt, E.; Ballesteros, A.; Bergner, F.; O'Donnell, I.; Efsing, P.; Hein, H.; Malerba, L.; Ortner, S.; Planman, T. (Eds.) *RPV Irradiation Embrittlement: NUGENIA Position on RPV Irradiation Embrittlement Issues Based on the Outcome of the EURATOM FP7 Project LONGLIFE*; NUGENIA, VTT Technical Research Center of Finland: Helsinki, Finland, 2015; p. 27.
12. Odette, G.; Lucas, G. Embrittlement of Nuclear Reactor Pressure Vessels. *JOM* **2001**, *53*, 18–22. [CrossRef]
13. Debarberis, L.; Kryukov, A.; Gillemont, F.; Acosta, B.; Sevini, F. Semi-mechanistic analytical model for radiation embrittlement and re-embrittlement data analysis. *Int. J. Press. Vessel. Pip.* **2005**, *82*, 195–200. [CrossRef]
14. Xiu, P.; Massey, C.P.; Kelsy Green, T.M.; Taller, S.; Isheim, D.; Sridharan, N.; Field, K.G. Microchemical evolution of irradiated additive-manufactured HT9. *J. Nucl. Mater* **2022**, *559*, 153410. [CrossRef]
15. Mamivand, M.; Wells, P.; Ke, H.; Shu, S.; Odette, G.R.; Morgan, D. CuMnNiSi precipitate evolution in irradiated reactor pressure vessel steels: Integrated Cluster Dynamics and experiments. *Acta Mater.* **2019**, *180*, 199–217. [CrossRef]
16. International Atomic Energy Agency. *Integrity of Reactor Pressure Vessels in Nuclear Power Plants: Assessment of Irradiation Embrittlement Effects in Reactor Pressure Vessel Steels*; International Atomic Energy Agency: Vienna, Austria, 2009; p. 144. ISBN 978-92-0-101709-3.
17. Nanstad, R.K.; Server, W.L. *Reactor Pressure Vessel Task of Light Water Reactor Sustainability Program: Initial Assessment of Thermal Annealing Needs and Challenges*; Oak Ridge National Laboratory: Oak Ridge, TN, USA, 2011.
18. Brumovsky, M.; Ahlstrand, R.; Brynda, J.; Debarberis, L.; Kohopaa, J.; Kryukov, A.; Server, W. Annealing and Re-Embrittlement of Reactor Pressure Vessel Materials. AMES Report N.19, JRC: Petten, Netherland. 2008. Available online: https://core.ac.uk/download/pdf/38614107.pdf (accessed on 20 January 2022).
19. Server, W.; Sokolov, M. Thermal Annealing of Reactor Pressure Vessels Is a Needed Mitigation Option. IAEA: Vienna, Austria. 2013. Available online: http://www.iaea.org/NuclearPower/Downloadable/Meetings/2013/2013-11-05-11-08-TM-NPE/38.Server_USA.pdf (accessed on 20 January 2022).
20. Brumovsky, M. *Embrittlement of Reactor Pressure Vessels (RPVs) in WWER-Type Reactors. Irradiation Embrittlement of Reactor Pressure Vessels (RPVs) in Nuclear Power Plants*; Woodhead Publishing: Sawston, UK, 2014; ISBN 9780857096470.
21. Slugen, V. *Safety of VVER-440 Reactors – Barriers Against Fission Products Release*; Springer: London, UK, 2011; ISBN 978-1-4471-6114-1.
22. Motter, F. *Low-Temperature Annealing of the BR-3 Reactor Vessel, NUREG/CP-0058-Vol.4*; U.S. Nuclear Regulatory Commission: Rockville, MD, USA, 1985; pp. 144–175.
23. Fabry, A. The BR3 Pressure Vessel Anneal: Lessons and Perspective. In Proceedings of the DOE/SNL/EPRI Workshop on Reactor Pressure Vessel Thermal Annealing, Albuquerque, NM, USA, 17–18 February 1994; pp. 5.1–5.33.
24. Mager, T. *Thermal Annealing of an Embrittled Reactor Pressure Vessel: Feasibility and Methodology*; EPRI NP-6113; Electric Power Research Institute: Washington, DC, USA, 1989.

25. Nanstad, R.K.; Tipping, P.; Kalkhof, R.D.; Sokolov, M.A. Effects of Irradiation and Post-Annealing Reirradiation on Reactor Pressure Vessel Steel Heat JRQ. In *The Effects of Radiation on Materials: 21st International Symposium, ASTM STP 1447*; Grossbeck, M.L., Allen, T.R., Lott, R.G., Kumar, A.S., Eds.; ASTM International: West Conshohocken, PA, USA, 2004; pp. 149–163.
26. Bergner, F.; Hernandez-Mayoral, M.; Heintze, C.; Konstantinovic, M.J.; Malerba, L.; Pareige, C. TEM Observation of Loops Decorating Dislocations and Resulting Source Hardening of Neutron-Irradiated Fe-Cr Alloys. *Metals* **2020**, *10*, 147. [CrossRef]
27. Konstantinovic, M.J.; Van Renterghem, W.; Matijasevic, M.; Minov, B.; Lambrecht, M.; Toyama, T.; Chiapetto, M.; Malerba, L. Mechanical and microstructural properties of neutron irradiated Fe–Cr–C alloys. *Phys. Status Solidi* **2016**, *213*, 2988–2994. [CrossRef]
28. Shimodaira, M.; Toyama, T.; Yoshida, K.; Inoue, K.; Ebisawa, N.; Tomura, K.; Yoshiie, T.; Konstantinovi, M.J.; Gérard, R.; Nagai, Y. Contribution of irradiation-induced defects to hardening of a low-copper reactor pressure vessel steel. *Acta Mater.* **2018**, *155*, 402–409. [CrossRef]
29. Amayev, A.D.; Kryukov, A.M.; Sokov, M.A. Recovery of the Transition Temperature of Irradiated WWER-440 Vessel Metal by Annealing. In *Radiation Embrittlement of Nuclear Pressure Vessel Steels: An International Review*; Steele, L.E., Ed.; ASTM International: West Conshohocken, PA, USA, 1993; Volume 4, pp. 369–379.
30. Alekseenko, N.N.; Amaev, A.; Gorynin, I.; Nikolaev, V.A. *Radiation Damage of Nuclear Power Plant Pressure Vessel Steels*; American Nuclear Society: La Grange Park, IL, USA, 1997; p. 282. ISBN 978-0-89448-564-0.
31. Was, G.S.; Jiao, Z.; Getto, E.; Sun, K.; Monterrosa, A.M.; Maloy, S.A.; Anderoglue, O.; Sencer, B.H.; Hackett, M. Emulation of reactor irradiation damage using ion beams. *Scr. Mater.* **2014**, *88*, 33–36. [CrossRef]
32. Was, G.S. Challenges to the use of ion irradiation for emulating reactor irradiation. *J. Mater. Res.* **2015**, *30*, 1–25. [CrossRef]
33. Was, G.S.; Allen, T.R. Radiation damage from different particle types. In *Radiation Effects in Solid*; NATO Science Series II: Mathematics, Physics and Chemistry; Sickafus, K.E., Kotomin, E.A., Uberuaga, B.P., Eds.; Springer: Berlin/Heidelberg, Germany, 2007; p. 235. ISSN 1568-2609.
34. Ghoneim, M.M.; Hammad, F.H. Pressure vessel steels: Influence of chemical composition on irradiation sensitivity. *Int. J. Press. Vess. Pip.* **1997**, *74*, 189–198. [CrossRef]
35. Kohopaa, J.; Ahlstrand, R. Re-embrittlement behaviour of VVER-440 reactor pressure vessel weld material after annealing. *Int. J. Press. Vess. Pip.* **2000**, *76*, 575–584. [CrossRef]
36. Debarberis, L.; von Estorff, U.; Crutzen, S.; Beers, M.; Stamm, H.; de Vries, M.I.; Tjoa, G.L. LYRA and other projects on RPV steel embrittlement: Study and mitigation of the AMES Network. *Nucl. Eng. Des.* **2000**, *195*, 217–226. [CrossRef]
37. Uniter States Nuclear Regulatory Commission. *Effects of Residual Elements on Predicted Radiation Damage to Reactor Vessel Materials*; U.S. NRC Regulatory Guide 1.99, Rev.1; Uniter States Nuclear Regulatory Commission: Washinghton, DC, USA, 1977; p. 7.
38. Suzuki, K. Reactor pressure vessel materials. In *Neutron Irradiation Effects in Reactor Pressure Vessel Steels and Weldments*; International Atomic Energy Agency: Vienna, Austria, 1998; pp. 70–164.
39. Grosse, M.; Denner, V.; Böhmert, J.; Mathon, M.H. Irradiation-induced structural changes in surveillance material of VVER 440-type weld metal. *J. Nucl. Mater.* **2000**, *277*, 280–287. [CrossRef]
40. Koutsky, J.; Kocik, J. *Radiaton Damage of Structural Materials*; Academy of Sciences of the Czech Republic: Prague, Czech Republic, 1994; p. 361.
41. Puska, M.J.; Nieminen, R.M. Theory of positrons in solids and on solid surfaces. *Rev. Mod. Phys.* **1994**, *66*, 841–897. [CrossRef]
42. Martin, J.W.; Paetsch, R. Interaction of positrons and dislocations. *J. Phys. F Met. Phys.* **1972**, *2*, 997–1008. [CrossRef]
43. Häkkinen, H.; Mäakinen, S.; Manninen, M. Edge dislocations in fcc metals: Microscopic calculations of core structure and positron states in Al and Cu. *Phys. Rev. B* **1990**, *41*, 12441–12453. [CrossRef]
44. Petersen, K.; Repin, I.A.; Trumpy, G. Positron lifetime analysis of dislocations arising from tensile strain. *J. Phys. Condens. Matter.* **1996**, *8*, 2815–2822. [CrossRef]
45. Smedskjaer, L.C.; Manninen, M.; Fluss, M.J. An alternative interpretation of positron annihilation in dislocations. *J. Phys. F Met. Phys.* **1980**, *10*, 2237–2249. [CrossRef]
46. Kamimura, Y.; Tsutsumi, T.; Kuramoto, E. Calculations of Positron Lifetimes in a Jog and Vacancies on an Edge-Dislocation Line in Fe. *Phys. Rev. B* **1995**, *52*, 879–885. [CrossRef]
47. Petrov, L.; Nankov, N.; Popov, E.; Troev, T. Positron Life Time Calculations of Defect in α-Iron Containing Hydrogen. In *AIP Conference Proceedings*; American Institute of Physics: College Park, MD, USA, 2008; Volume 996, pp. 177–182.
48. Kamimura, Y.; Tsutsumi, T.; Kuramoto, E. Influence of Dislocations on Positron Lifetime in Iron. *J. Phys. Soc. Jpn.* **1997**, *66*, 3090–3096. [CrossRef]
49. Park, Y.K.; Waber, J.T.; Meshii, M.; Snead, C.L.; Park, C.G. Dislocation studies on deformed single crystals of high-purity iron using positron annihilation: Determination of dislocation densities. *Phys. Rev. B* **1986**, *34*, 823. [CrossRef]
50. Hodges, C.H. Positron diffusion and trapping at vacancies. *J. Phys. F Met. Phys.* **1974**, *4*, L230. [CrossRef]
51. Dupasquier, A.; Romero, R.; Somoza, A. Positron trapping at grain boundaries. *Phys. Rev. B* **1993**, *48*, 9235. [CrossRef]
52. Selim, F.A. Positron annihilation spectroscopy of defects in nuclear and irradiated materials–A review. *Mater. Charact.* **2021**, *174*, 110952. [CrossRef]
53. Phythian, W.J.; English, C.A. Microstructural evolution in reactor pressure vessel steels. *J. Nucl. Mater.* **1993**, *205*, 162–177. [CrossRef]

54. Slugen, V. Microstructural Analysis of Nuclear Reactor Pressure Vessel Steels. In *Mössbauer Spectroscopy in Material Science*; Miglierini, M., Petridis, D., Eds.; Kluwer Academic Publishers: Dordrecht, The Netherlands, 1999; pp. 119–130.
55. De Bakker, P.; Slugen, V.; De Grave, E.; Van Walle, E.; Fabry, A. Differences between eastern and western-type nuclear reactor pressure vessel steels as probed by Mössbauer spectroscopy. *Hyperfine Interact.* **1997**, *110*, 11–16. [CrossRef]
56. Brauer, G.; Liszkay, L.; Molnar, B.; Krause, R. Microstructural aspects of neutron embrittlement of reactor pressure vessel steels - A view from positron annihilation spectroscopy. *Nucl. Eng. Desg.* **1991**, *127*, 47–68. [CrossRef]
57. Pareja, R.; De Diego, N.; De La Cruz, R.M.; Rio, J.D. Postirradiation recovery of a reactor pressure vessel steel investigated by positron annihilation and microhardness measurements. *Nucl. Technol.* **1993**, *104*, 52–63. [CrossRef]
58. Slugen, V.; Zeman, A.; Petriska, M.; Krsjak, V. Positron study of radiation embrittlement of steels used in water cooled, water moderated energy reactors. *Appl. Surf. Sci.* **2006**, *252*, 3309–3315. [CrossRef]
59. Valo, M.; Krause, R.; Saarinen, K.; Hautojärvi, P.; Hawthorne, R. Irradiation Response and Annealing Behaviour of Pressure Vessel Model Steels and Iron Ternary Alloys Measured with Positron Techniques. In *Effects of Radiation on Materials: 15th International Symposium*; ASTM STP, 1125; Stoller, R.E., Kumar, A.S., Gelles, D.S., Eds.; American Society for Testing and Materials: West Conshohocken, PA, USA, 1992; pp. 172–185.
60. Hartley, J.H.; Howell, R.H.; Asoka-Kumar, P.; Sterne, P.A.; Akers, D.; Denison, A. Positron annihilation studies of fatigue in 304 stainless steel. *Appl. Surf. Sci.* **1999**, *149*, 204–206. [CrossRef]
61. Becvar, F.; Cizek, J.; Lestak, L.; Novotny, I.; Prochazka, I.; Sebesta, F. A high-resolution BaF_2 positron-lifetime spectrometer and experience with its long-term exploitation. *Nucl. Instr. Meth. A* **2000**, *443*, 557–577. [CrossRef]
62. Miller, M.K.; Russel, K.F.; Kocik, J.; Keilova, E. Embrittlement of low copper VVER 440 surveillance samples neutron-irradiated to high fluences. *J. Nucl. Mater.* **2000**, *282*, 83–88. [CrossRef]
63. Cizek, J.; Becvar, F.; Prochazka, I. Three-detector setup for positron-lifetime spectroscopy of solids containing ^{60}Co radionuclide. *Nucl. Instr. Meth. A* **2000**, *450*, 325–337. [CrossRef]
64. Cizek, J.; Prochazka, I.; Kocik, J.; Keilova, E. Positron Lifetime Study of Reactor Pressure Vessel Steels. *Phys. Stat. Sol. (A)* **2000**, *178*, 651–662. [CrossRef]
65. Van Hoorebeke, L.; Fabry, A.; van Walle, E.; Van de Velde, J.; Segers, D.; Dorikens-Vanpraet, L. A three-detector positron lifetime setup suited for measurements on irradiated steels. *Nucl. Instr. Meth. A* **1996**, *371*, 566–571. [CrossRef]
66. Ghazi-Wakili, K.; Zimmermann, U.; Brunner, J.; Tipping, P.; Waeber, W.B.; Heinrich, F. Positron Annihilation Studies on Neutron Irradiated Pressure Vessel Steels. *Phys. Stat. Sol. (A)* **1987**, *102*, 153–163. [CrossRef]
67. Slugen, V.; Segers, D.; De Bakker, P.M.A.; DeGrave, E.; Magula, V.; Van Hoecke, T.; Van Waeyenberge, B. Annealing behaviour of reactor pressure-vessel steels studied by positron-annihilation spectroscopy, Mössbauer spectroscopy and transmission electron microscopy. *J. Nucl. Mater.* **1999**, *274*, 273–286. [CrossRef]
68. Slugen, V.; Magula, V. The micro structural study of 15Kh2MFA and 15Kh2NMFA reactor pressure vessel steels using positron-annihilation spectroscopy, mössbauer spectroscopy and transmission electron microscopy. *Nucl. Eng. Desg.* **1998**, *186*, 323–342. [CrossRef]
69. Slugen, V.; De Grave, E.; Segers, D. Microstructural study of thermally treated reactor pressure vessel steels using spectroscopic methods. *Int. J. Nucl. Energy Sci. Technol.* **2004**, *1*, 20–32. [CrossRef]
70. Slugen, V.; Hascik, J.; Gröne, R.; Bartik, P.; Zeman, A.; Kögel, G.; Sperr, P.; Triftshäuser, W. Investigation of Reactor Steels. *Mater. Sci. Forum* **2001**, *363–365*, 47–51. [CrossRef]
71. Lambrecht, M.; Almazouzi, A. Positron annihilation study of neutron irradiated model alloys and of a reactor pressure vessel steel. *J. Nucl. Mater.* **2009**, *385*, 334–338. [CrossRef]
72. Puska, M.J.; Sob, M.; Brauer, G.; Korhonen, T. First-principles calculation of positron lifetimes and affinities in perfect and imperfect transition-metal carbides and nitrides. *Phys. Rev. B* **1994**, *49*, 10947. [CrossRef]
73. Slugen, V.; Hein, H.; Sojak, S.; Veternikova, J.; Petriska, M.; Sabelova, V.; Pavuk, M.; Hinca, R.; Stacho, M. Evaluation of the Reactor Pressure Vessel Steels by Positron Annihilation. *J. Nucl. Mater.* **2013**, *442*, 499–506. [CrossRef]
74. Pecko, S.; Sojak, S.; Slugen, V. Comparative study of irradiated and hydrogen implantation damaged German RPV steels from PAS point of view. *Appl. Surf. Sci.* **2014**, *312*, 172–175. [CrossRef]
75. Nagai, Y.; Tang, Z.; Hasegawa, M.; Kanai, T.; Saneyasu, M. Irradiation-induced Cu aggregations in Fe: An origin of embrittlement of reactor pressure vessel steels. *Phys. Rev. B* **2001**, *63*, 131110. [CrossRef]
76. Magula, V.; Janovec, J. Effect of short time high temperature annealing on kinetics of carbidic reactions in 2.7 Cr-O. 6Mo-0.3 V steel. *Ironmak. Steelmak.* **1994**, *21*, 64.
77. Kocik, J.; Keilova, E.; Cizek, J.; Prochazka, I. In Proceedings of the 9th International Conference on Metallurgy METAL 2000; Tanger Ltd.: Ostrava, Czech Republic, 2000. Paper No. 719.
78. Brandt, W. *Positron Annihilation*; Stewart, A.T., Roelling, L.O., Eds.; Academic Press: New York, NY, USA, 1967; p. 155.
79. Frank, W.; Seeger, A. Theoretical foundation and extension of the trapping model. *Appl. Phys.* **1974**, *3*, 61–66. [CrossRef]
80. Seeger, A. The study of defects in crystals by positron annihilation. *Appl. Phys.* **1974**, *4*, 183. [CrossRef]
81. Nieminen, R.N.; Laakkonen, J.; Hautojärvi, P.; Vehanen, A. Temperature dependence of positron trapping at voids in metals. *Phys. Rev. B* **1979**, *19*, 1397–1402. [CrossRef]
82. Frieze, W.E.; Lynn, K.G.; Welch, D.O. Positron trapping model including spatial diffusion of the positron. *Phys. Rev. B* **1985**, *31*, 15. [CrossRef]

83. Britton, D.T. Time-dependent diffusion and annihilation of positrons implanted in a semi-infinite medium. *J. Phys. Condens. Matte* **1991**, *3*, 681. [CrossRef]
84. Kögel, G. Positron diffusion in solids and the reconstruction of inhomogeneous defect distributions from lifetime measurements. *Appl. Phys. A* **1996**, *63*, 227. [CrossRef]
85. Sperr, P.; Egger, W.; Kögel, G.; Dollinger, G.; Hugenschmedt, C.; Repper, R.; Piochacz, C. Status of the pulsed low energy positron beam system (PLEPS) at the Munich Research Reactor FRM-II. *Appl. Surf. Sci.* **2008**, *255*, 35–38. [CrossRef]
86. Hugenschmidt, C. Positron in surface physics. *Surf. Sci. Rep.* **2016**, *71*, 547–594. [CrossRef]
87. Hugenschmidt, C.; Löwe, B.; Mayer, J.; Piochacz, C.; Pikart, P.; Repper, R.; Stadlbauer, M.; Scheckenbach, K. Unprecedented intensity of a low-energy positron beam. *Nucl. Instrum. Meth. A* **2008**, *593*, 616–618. [CrossRef]
88. Steele, L.E. *Radiation Embrittlement of Nuclear Reactor Pressure Vessel Steels: An International Review (Fourth Volume)*; ASTM International: West Conshohocken, PA, USA, 1993; ISBN 0-8031-1478-8.
89. Noga, P.; Dobrovodsky, J.; Vana, D.; Beno, M.; Zavacka, A.; Muska, M.; Halgas, R.; Minarik, S.; Riedlmajer, R. A new ion-beam laboratory for materials research at the Slovak University of Technology. *Nucl. Instr. Meth. B* **2017**, *409*, 264–267. [CrossRef]
90. Edwardson, C. Positron Studies of Defects in Thin Films and Semiconductors. Ph.D. Thesis, University of Bath, Bath, UK, 2013.
91. Krause-Rehberge, R.; Leipner, S.H. *Positron Annihilation in Semiconductors*; Springer: Berlin, Germany, 1998; ISBN 3-540-64371-0.
92. Eldrup, M. Positron Methods for the Study of Defects in Bulk Materials. *J. Phys. IV* **1995**, *5*, C1-93–C1-109. [CrossRef]
93. Saro, M.; Kršjak, V.; Lauko, R.; Slugeň, V. Application of Na-22 positron source to the investigation of ion-implanted iron samples. In *AIP Conference Proceedings*; AIP Publishing LLC: Melville, NY, USA, 2019; Volume 2182.
94. Petriska, M.; Sabelova, V.; Slugeň, V.; Sojak, S.; Stacho, M.; Veterníková, J. Digital Coincidence Doppler Broadening setup at FEI STU. *Phys. Procedia* **2012**, *35*, 117–121. [CrossRef]
95. Kuriplach, J.; Melikhova, O.; Domain, C.; Becquart, C.S.; Kulikov, D.; Malerba, L.; Hou, M.; Almazouzi, A.; Duque, C.A.; Morales, A.L. Vacancy-solute complexes and their clusters in iron. *Appl. Surf. Sci.* **2006**, *252*, 3303–3308. [CrossRef]
96. Dryzek, J.; Horodek, P. Positron Annihilation Studies of the Near-Surface Regions of Niobium before and after Wear Treatment. *Tribol. Lett.* **2017**, *65*, 117. [CrossRef]
97. Wu, Y.C.; Chen, Y.Q.; Wang, B.; Wang, S.J.; Jean, Y.C.; Suzuki, R.; Ohdaira, T. Slow positron beam study of corrosion-related defects in pure iron. *Appl. Surf. Sci.* **2006**, *252*, 3274–3277. [CrossRef]
98. Sato, K.; Kinomura, A.; Omura, T.; Xu, Q.; Yoshiie, T.; Kasada, R.; Kimura, A.; Morishita, K. Positron annihilation lifetime measurements of He-ion-irradiated Fe using pulsed positron beam. In *Journal of Physics: Conference Series, Volume 262, 12th International Workshop on Slow Positron Beam Techniques (SLOPOS12) 1–6 August 2010, Magnetic Island, North Queensland, Australia*; IOP Publishing: Bristol, UK, 2011; p. 012053.
99. Troev, T.; Popov, E.; Staikov, P.; Nankov, N. Positron lifetime studies of defects in α-Fe containing helium. *Phys. Status Solidi C* **2009**, *6*, 2373–2375. [CrossRef]
100. Horodek, P.; Skuratov, V.A. Variable energy positron beam studies of defects in heavy ion implanted palladium. *Surf. Coat. Technol.* **2016**, 296. [CrossRef]
101. Kansy, J. Microcomputer program for analysis of positron annihilation lifetime spectra. *Nucl. Instr. Meth. Phys. Res. A* **1996**, *374*, 235–244. [CrossRef]
102. Petersen, K. *Crystal Defects Studied by Positrons*; Politeknik Forlag: Lyngby, Denmark, 1978; ISBN 8750204882.
103. Zhang, J.; Liu, F.; Cheng, G.; Shang, J.; Liu, J.; Cao, S.; Liu, Z. Electron structure and vacancy properties and Al-substitution dependence of the positron lifetime in Y1:2:3 superconducting ceramics. *Phys. Lett. A* **1995**, *201*, 70–76. [CrossRef]
104. Jansson, V.; Malerba, L. Simulation of the nanostructure evolution under irradiation in Fe–C alloys. *J. Nucl. Mater.* **2013**, *443*, 274–285. [CrossRef]
105. Jansson, V.; Chiapetto, M.; Malerba, L. The nanostructure evolution in Fe–C systems under irradiation at 560 K. *J. Nucl. Mater.* **2013**, *442*, 341–349. [CrossRef]
106. Anento, N.; Serra, A. Carbon–vacancy complexes as traps for self-interstitial clusters in Fe–C alloys. *J. Nucl. Mater.* **2013**, *44*, 236–242. [CrossRef]
107. Terentyev, D.; Anento, N.; Serra, A.; Jansson, V.; Khater, H.; Bonny, G. Interaction of carbon with vacancy and self-interstitial atom clusters in a-iron studied using metallic–covalent interatomic potential. *J. Nucl. Mater.* **2011**, *408*, 272–284. [CrossRef]
108. Shibamoto, H.; Kimura, A.; Hasegawa, M.; Matsui, H.; Yamaguchi, S. Effects of Proton Irradiation on Reactor Pressure Vessel Steel and Its Model Alloys. *J. ASTM Int.* **2005**, *2*, 9. [CrossRef]
109. Dai, G.H.; Li, X.H.; Moser, P.; Moya, G.; Van Duysen, J.C. Defect Recovery in α -Fe e^--Irradiated at 300 K. *Acta Phys. Pol. A* **1993**, *3*, 277–286. [CrossRef]
110. Iwai, T. A positron beam Doppler broadening analysis of formation and recovery of defects produced by ion irradiation in Fe−C−Cu alloys. *Radiat. Eff. Defects Solids* **2013**, *168*, 308–315. [CrossRef]
111. Takaki, S.; Fuss, J.; Kuglers, H.; Dedek, U.; Schultz, H. The resistivity recovery of high purity and carbon doped iron following low temperature electron irradiation. *Radiat. Eff. Defects Solids* **1983**, *79*, 87–122. [CrossRef]
112. Hari Babu, S.; Amarendra, G.; Rajaraman, R.; Sundar, C.S. Microstructural Characterization of Ferritic/Martensitic Steels by Positron Annihilation Spectroscopy. *J. Nucl. Mater.* **2013**, *432*, 012010. [CrossRef]
113. Ilola, R.; Nadutov, V.; Valo, M.; Hanninen, H. On irradiation embrittlement and recovery annealing mechanisms of Cr–Mo–V type pressure vessel steels. *J. Nucl. Mater.* **2002**, *302*, 185–192. [CrossRef]

114. Jones, W.B.; Hills, C.R.; Polonis, D.H. Microstructural evolution of modified 9Cr-1Mo steel. *Metall. Trans. A* **1991**, *22*, 1049–1058. [CrossRef]
115. Slugen, V.; Kogel, G.; Sperr, P.; Triftshauser, W. Positron annihilation studies of neutron irradiated and thermally treated reactor pressure vessel steels. *J. Nucl. Mater.* **2002**, *302*, 89–95. [CrossRef]
116. Slugen, V.; Zeman, A.; Lipka, J.; Debarberis, L. Positron annihilation and Mossbauer spectroscopy applied to WWER-1000 RPV steels in the frame of IAEA High Ni Co-ordinated Research Programme. *NDTE Int.* **2004**, *37*, 651–661. [CrossRef]
117. Abe, F. Precipitate design for creep strengthening of 9% Cr tempered martensitic steel for ultra-supercritical power plants. *Sci. Technol. Adv. Mater.* **2008**, *9*, 013002. [CrossRef]
118. Zeman, A.; Debarberis, L.; Kocik, J.; Slugen, V.; Keilova, E. Microstructural analysis of candidate steels pre-selected for new advanced reactor systems. *J. Nucl. Mater.* **2007**, *362*, 259. [CrossRef]
119. Zeman, A.; Debarberis, L.; Slugen, V.; Acosta, B. Assessment of the correlation between mechanical testing and positron annihilation outcomes for RPV model alloys. *Appl. Surf. Sci.* **2005**, *252*, 3290. [CrossRef]
120. Dai, Y.; Krsjak, V.; Kuksenko, V.; Schaeublin, R. Microstructural changes of ferritic/martensitic steels after irradiation in spallation target environments. *J. Nucl. Mater.* **2018**, *511*, 508–522. [CrossRef]
121. David, A.; Kögel, G.; Sperr, P.; Triftshäuser, W. Lifetime Measurements with a Scanning Positron Microscope. *Phys. Rev. Lett.* **2001**, *87*, 067402. [CrossRef]

materials

MDPI

Article

Lattice Defects and Exfoliation Efficiency of 6H-SiC via H_2^+ Implantation at Elevated Temperature

Tao Wang [1,*], Zhen Yang [2,*], Bingsheng Li [3,*], Shuai Xu [3], Qing Liao [3], Fangfang Ge [4], Tongmin Zhang [5] and Jun Li [5]

1 Institute of Fluid Physics, China Academy of Engineering Physics, Mianyang 621900, China
2 Sino-French Institute of Nuclear Engineering and Technology, Sun Yat-Sen University, Zhuhai 519082, China
3 State Key Laboratory for Environment-Friendly Energy Materials, Southwest University of Science and Technology, Mianyang 621010, China; xushuai@swust.edu.cn (S.X.); Liaoq@swust.edu.cn (Q.L.)
4 Ningbo Institute of Materials Technology and Engineering, Chinese Academy of Sciences, Ningbo 315201, China; gefangfang@nimte.ac.cn
5 Institute of Modern Physics, Chinese Academy of Science, Lanzhou 730000, China; zhangtm@impcas.ac.cn (T.Z.); lijun@impcas.ac.cn (J.L.)
* Correspondence: wangtaoxtc@gmail.com (T.W.); yangzh97@mail.sysu.edu.cn (Z.Y.); b.s.li@swust.edu.cn (B.L.)

Received: 20 October 2020; Accepted: 8 December 2020; Published: 15 December 2020

Abstract: Silicon carbide (SiC) is an important material used in semiconductor industries and nuclear power plants. SiC wafer implanted with H ions can be cleaved inside the damaged layer after annealing, in order to facilitate the transfer of a thin SiC slice to a handling wafer. This process is known as "ion-cut" or "Smart-Cut". It is worth investigating the exfoliation efficiency and residual lattice defects in H-implanted SiC before and after annealing. In the present paper, lattice damage in the 6H-SiC implanted by H_2^+ to a fluence of 5×10^{16} H_2^+/cm^2 at 450 and 900 °C was investigated by a combination of Raman spectroscopy and transmission electron microscopy. Different levels of damage caused by dynamic annealing were observed by Raman spectroscopy and transmission electron microscopy in the as-implanted sample. Atomic force microscopy and scanning white-light interferometry were used to observe the sample surface morphology. Surface blisters and exfoliations were observed in the sample implanted at 450 °C and then annealed at 1100 °C for 15 min, whereas surface blisters and exfoliation occurred in the sample implanted at 900 °C without further thermal treatment. This finding can be attributed to the increase in the internal pressure of platelets during high temperature implantation. The exfoliation efficiency, location, and roughness after exfoliation were investigated and possible reasons were discussed. This work provides a basis for further understanding and improving the high-efficiency "ion-cut" technology.

Keywords: 6H-SiC; H_2^+ implantation; exfoliation; microstructure

1. Introduction

Silicon carbide (SiC) is regarded as one of the most important wide-band gap semiconductors due to its excellent physical, electronic, and optical performances, i.e., a high melting temperature, a high strength, a high thermal conductivity, a large breakdown voltage, and a high electron mobility [1,2]. Much effort has been made to develop the potential applications of SiC devices, such as Schottky barrier diodes in next-generation, large-scale integrated circuits. Although SiC has more than 200 polytypes, the hexagonal 4H- and 6H-SiC are particularly promising due to their advanced physical properties.

To reduce the high cost of SiC wafers and improve SiC-devices, SiC-on-insulator (SiCOI) structures have been proposed because of their excellent performance, such as the low-power dissipation to save energy and the high radiation resistance to use in space [3]. Similar to many other semiconductors, SiCOI structures can be fabricated by "smart-cut" technology, which was first reported by Bruel [4] in 1995, to achieve silicon layer transfer for the fabrication of silicon-on-insulator (SOI) materials. The "Smart-Cut" technology contains three main processes, initially hydrogen or helium ion implantation with a fluence of the order of 10^{16} to 10^{17} cm^{-2} at room temperature, then wafer bonding to another rigid substrate (handling wafer) before thermal annealing, and finally fracture to achieve thin layer transfer at elevated temperatures [5–7]. The initiation and propagation of micro-cracks in H-implanted SiC play a critical role in exfoliation of the wafer surface. The formation of micro-cracks depends on the growth of platelets. These platelets are composed of vacancy-hydrogen compounds. The formation of vacancy-hydrogen compounds is due to the interaction between implantation-induced vacancies and implanted hydrogen. Therefore, it is critical to investigate the formation and growth of platelets in SiC implanted with H ions under different experimental conditions, such as the implantation fluence, temperature, and annealing treatment. It is well known that the growth of micro-cracks inside the SiC wafer can induce surface blisters when the SiC wafer is not bonded to a substrate, and the same activation energy between blister formation and layer splitting is argued by Tong et al. [8]; therefore, it is a convenient way to evaluate the smart-cut threshold condition via observation of surface blisters and exfoliation. Our recent study involved 6H-SiC implanted by 134 keV H_2^+ at room temperature [9]. The maximum exfoliation efficiency was achieved for the sample implanted with a fluence of 1.5×10^{16} H_2^+/cm^2 followed by 1100 °C annealing for 15 min. A further increase in implantation fluence was found to retard the exfoliation efficiency due to the negative effects of implantation-induced lattice damage on the growth of vacancy-hydrogen clusters, consistent with the report of Gregory et al. [10] that the threshold fluence for exfoliation in H-implanted 4H-SiC decreases with increasing implantation temperature (room temperature to 600 °C). Up to now, most of the published reports aimed at H-implanted SiC were concerned with implantation at a low temperature and then annealing at a high temperature [11–17]. The exfoliation effect in H-implanted SiC without the annealing treatment was, to our knowledge, not investigated. Many open questions are concerned with the nature of the H implantation-induced defects and their influence on micro-crack growth. In this paper, we studied the exfoliation efficiency of 6H-SiC implanted at 450 °C and subsequently annealed at 1100 °C for 15 min, compared with 6H-SiC implanted at 900 °C without annealing.

2. Experimental Process

For the experiments to study the exfoliation efficiency of SiC as a function of implantation temperature, bulk SiC samples, 6H polytype $<0001>_{Si}$ orientation, purchased as research grade material from HF-Kejing Company, Heifei, China, were implanted with 194 keV H_2^+ to a fluence of 5×10^{16} H_2^+/cm^2 at 450 and 900 °C. Hydrogen implantation experiments were performed on a 320 kV high-voltage platform equipped with ECR (Electron Cyclotron Resonance) ion sources in the Institute of Modern Physics, Chinese Academy of Sciences (CAS). The beam was rastered using an electrostatic scanner with fixed frequencies of 993 and 990 Hz in horizontal and vertical directions, respectively, to provide uniform ion fluence across the sample. The ion fluence was in-situ measured using a Faraday cup assembly in front of the sample. The beam flux was kept at 2.3×10^{13} ions/cm^2 s. The implantation temperature was measured by a thermocouple, and the deviation of the implantation temperature was less than 1 °C. The wafers were tilted 7–8° from the direction of normal incidence during the implantation. According to the Stopping and Range of Ions in Matter (SRIM-2013) [18], the expected H peak concentration was approximately 11 at.% at 576 nm below the sample surface, as shown in Figure 1. To observe exfoliation on the surface of H-implanted 6H-SiC at 450 °C, thermal annealing at 1100 °C for 15 min in air atmosphere was performed.

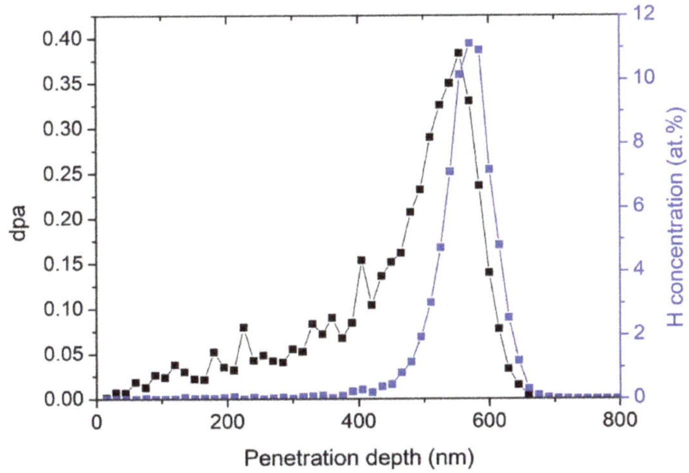

Figure 1. Depth distributions of the displacements per atom (dpa) and the projected range of 194 keV H_2^+-implanted 6H-SiC to a fluence of 5×10^{16} H_2^+/cm^2 simulated using the SRIM-2013 code (density of 3.21 g/cm^3 and displacement energies of C = 20 eV and Si = 35 eV).

Lattice damage before and after annealing was investigated by Raman spectroscopy and transmission electron microscopy (TEM) using a Tecnai G20 operated at 200 kV. Confocal Raman spectra were recorded at room temperature in a $\bar{z}(xx)z$ backscattering geometry using an HR-800 spectrometer from France. The 532 nm line of an argon ion laser was focused on a 1×1 μm^2 spot and collected through a 50× objective lens. A 100 μm confocal pinhole diameter was used, and 600 lines/mm grating were performed. The acquisition time for each spectrum was 30 s for one accumulation. The spectra were measured ranging from 150 to 1800 cm^{-1}. A double tilt goniometer stage was used, in order to tilt the TEM sample to satisfy different diffraction vectors. The lattice defects were detected by weak-beam dark-field (WBDF) with $(g, 3g)$, $g = 0002$ and $g = \bar{2}110$ near $z = 01\bar{1}0$, where g is the diffraction vector and z is the zone axis. To study the depth distribution of implantation-induced defects, cross-sectional samples were prepared. The fabrication process of the cross-sectional transmission electron microscopy (XTEM) samples was described as follows. Initially, XTEM samples were prepared by mechanical thinning up to approximately 30 μm in thickness, followed by ion milling with Ar ions in two steps. In the first step, the ion milling energy was 5 kV with a glancing angle of ±5° until optically controlled perforation occurred in the middle of the XTEM sample. In the second step, ion milling energy decreased to 2 kV with a glancing angle of ± 3° for 1 h to minimize radiation damage induced by the Ar ions [14,15]. The surface morphology was measured by scanning white-light interferometry (SWLI), and surface roughness after exfoliation was measured by atomic force microscopy (AFM).

3. Results

Figure 2 presents the Raman spectra of the 194 keV H_2^+-implanted 6H-SiC to a fluence of 5×10^{16} H_2/cm^2 at 450 and 900 °C. In the as-grown 6H-SiC, some Raman scattering peaks were clearly visible. Nakashima and Harima [19] investigated the Raman scattering of SiC crystals, and they found that the Raman-active models of the wurtzite structure were the A_1, E_1, and E_2 modes. In addition, the A_1 and E_2 phonon modes can be split into transverse acoustic (TA) and optical (TO), as well as longitudinal acoustic (LA) and optical (LO) modes. The first-order Raman bands assigned to E_2(TO) at 767 and 789 cm^{-1} and A_1(LO) at 967 cm^{-1} were observed [20]. Besides the first-order Raman bands, second-order Raman bands attributed to E_2(TA) at 146 and 150 cm^{-1}, E_2(TA) at 266 cm^{-1}, and A_1(LA) at 504 and 513 cm^{-1} [20]. The Raman active located in the 1500–1750 cm^{-1} region can be attributed to optical branching [20]. The strong intensity of the second-order Raman bands indicates the good quality of the wafer. After H_2^+ ion implantation, the intensities of the first-order Raman bands and the second-order Raman bands decreased. This finding can be assigned to the increase in the optical absorption coefficient of 6H-SiC after H implantation [21,22]. It is a simple method to evaluate the lattice disorder by means of the change in Raman scattering intensity. The Raman scattering of A_1(LO) is enlarged and shown in Figure 2b. Compared with the H_2^+-implanted 6H-SiC at 900 °C, in H_2^+-implanted 6H-SiC at 450 °C, the Raman scattering decreased more significantly. Moreover, the asymmetric broadening of the A_1(LO) peak can be observed. In detail, the left tail of the A_1(LO) peak lifted after H_2^+ ion implantation. The intensity of the asymmetry can be expressed as $\Delta\tau = (I_{left} - I_{right})/I_{right}$, where I is the intensity of the Raman scattering baseline; $\Delta\tau$ equaled 76% and 30% at the 450 °C and 900 °C implantation, respectively. The asymmetric broadening of the A_1(LO) peak can be accounted for by a "spatial correlation" model where implantation-induced defects can induce q-vector relaxation [23,24]. The more lattice defects in the wafer, the stronger the asymmetric broadening that can be formed [25]. Therefore, it is reasonable to assume that the number density of lattice defects formed at 450 °C implantation is larger than that of the sample implanted at 900 °C. It is easily explained that the dynamic annealing increases with increasing implantation temperature.

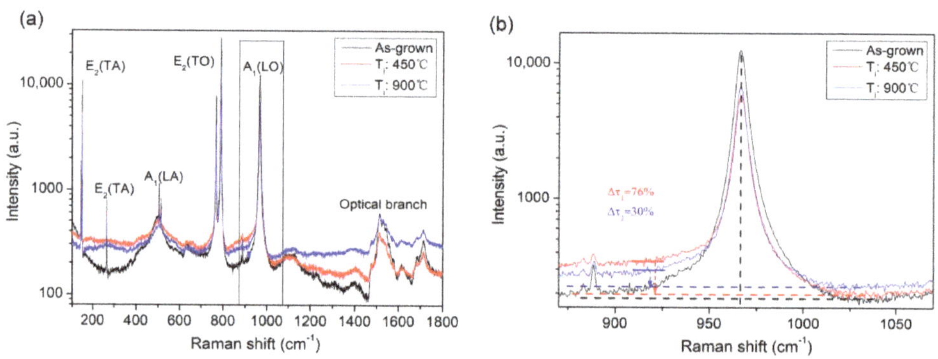

Figure 2. Raman spectra of (**a**) 6H-SiC implanted by H_2^+-ion at 450 °C and 900 °C showing first-order (E_2(TO) and A_1(LO)) and second-order (E_2(TA), E_1(TA), A_1(LA) and optical branch) peaks, compared with the as-grown 6H-SiC, (**b**) enlarged A_1(LO) peak shown in figure (**a**), where the left tail of the peak lifted after ion implantation. The intensity of the asymmetry decreased with increasing implantation temperature.

Figure 3 shows the surface morphology of the 194 keV H_2^+-implanted SiC by means of the scanning white-light interferometry method. In the 6H-SiC implanted with H_2^+ ions at 450 °C after 1100 °C annealing for 15 min, surface exfoliation was clearly observed in Figure 3a,d. Surface blisters were observed in the two-dimensional profile shown in Figure 3b,f. Exfoliation depth and size were analyzed by a contour curve, as shown in Figure 3c,g. It can be seen that most of the exfoliation depth is near 1.0 μm for the H_2^+-implanted 6H-SiC at 450 °C followed by 1100 °C annealing, while it is near 0.8 μm for the H_2^+-implanted 6H-SiC at 900 °C It should be noted that the exfoliation depth observed by scanning white light interferometry is not exact due to lattice swelling induced by surface blisters. Moreover, the size of the exfoliation zone is in the range of 10 to 40 μm for the H_2^+-implanted 6H-SiC at 450 °C followed by 1100 °C annealing, while it is in the range of 20 to 100 μm for the 6H-SiC implanted with H_2^+ ions at 900 °C. This result demonstrates that the exfoliation efficiency of the 6H-SiC implanted with H_2^+ ions at 900 °C is higher than that of the 6H-SiC implanted with H_2^+ ions at 450 °C followed by 1100 °C annealing.

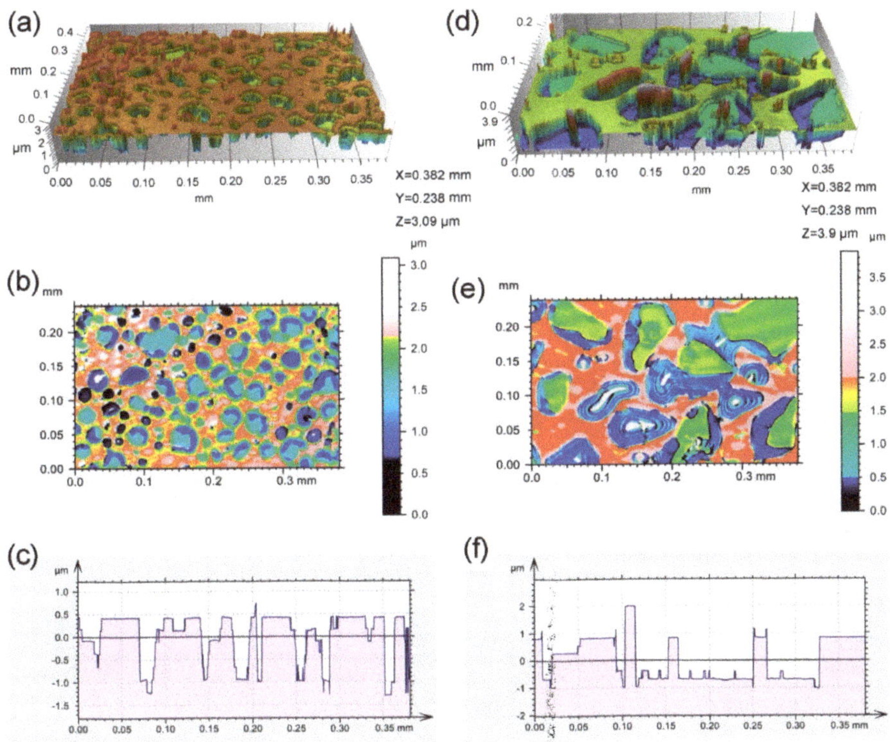

Figure 3. (**a**,**d**) show surface morphology observed in the three-dimensional profile, (**b**,**d**) show the height distribution in a two-dimensional profile, and (**c**,**f**) show the contour curve obtained from figures (**b**,**d**), respectively. (**a**–**c**) are the 194 keV H_2^+-implanted 6H-SiC at 450 °C followed by 1100 °C annealing for 15 min. (**d**–**f**) are the 194 keV H_2^+-implanted 6H-SiC at 900 °C.

To carefully investigate the surface morphology after H_2^+-implantation into 6H-SiC, an AFM test was performed, and results are presented in Figure 4. It can be seen that the shape of the exfoliation zone is not regular, near an oval shape, as shown in Figure 4a, c. The formation of the exfoliation zone is due to the breakage of a surface blister when its inner stress exceeds the material fracture toughness [26–28]. The exfoliation zone is presented by a three-dimensional image, and the surface is not even, consisting of many hillocks. The values of the root-mean-square (RMS) roughness are 12.9 nm and 10.1 nm for the H_2^+-implanted 6H-SiC at 450 °C followed by 1100 °C annealing for 15 min and the H_2^+-implanted 6H-SiC at 900 °C, respectively. The decrease in RMS for the 6H-SiC implanted with H_2^+ ions at 900 °C is due to the fast growth of hydrogen-vacancy clusters, resulting in the increase in exfoliation efficiency, as observed by using the scanning white-light interferometry method (see Figure 3).

To explain the higher exfoliation efficiency in the sample implanted at 900 °C compared to the sample implanted at 450 °C and consequently annealed at 1100 °C, microstructures of lattice defects and microcracks were investigated by XTEM. Figure 5 presents the general view of the lattice defects formed in the 194 keV H_2^+-implanted 6H-SiC at 450 °C. It can be seen in Figure 5a that the damage band exhibiting a black contrast is located at a depth ranging from approximately 540 to 650 nm beneath the surface. According to SRIM-2013 simulation, the projected range of 194 keV H_2^+ implantation is 553 nm with a straggling range of 52 nm. This implies that the measured damage band is deeper than the simulated projected range. This result can be accounted for by lattice swelling due to dense interstitial-type defects produced by H_2^+ implantation. It is reasonable to expect the lattice swelling to be approximately 5%. To investigate lattice defects in the damage band, bright-field (BF) and WBDF observations with two different diffraction vectors were performed, as shown in Figure 5b–e. Due to the nano-scaled lattice defects, these lattice defects are easily distinguished under the WBDF condition. It can be seen that many bright spots were observed, and some bright spots have larger sizes at the bottom of the damage band compared with the front of the damage band. The width of the observed damage band is approximately 120 nm. In the front of the damage band, many point defect clusters were observed under $g = 0002$, but not at $g = \bar{2}110$. This result indicated that these point defect clusters are Frank loops with a Burgers vector of 1/2<0001>. At the bottom of the damage band, some large defect clusters were observed under $g = 0002$ and $g = \bar{2}110$ simultaneously. It is indicated that these defect clusters have a Burgers vector of 1/6<2203>. The distribution of the lattice atoms was measured by high-resolution TEM (HRTEM), and the observed lattice defects exhibited a black contrast due to Bragg diffraction, as shown in Figure 5f. It can be seen in Figure 5f,g that lattice fringes are seriously disturbed by H_2^+ implantation. Because the C and Si vacancies cannot migrate at 450 °C, most of the observed lattice defects are composed of interstitial atoms, such as C interstitials, which induce a significant lattice swelling at 450 °C implantation [29,30].

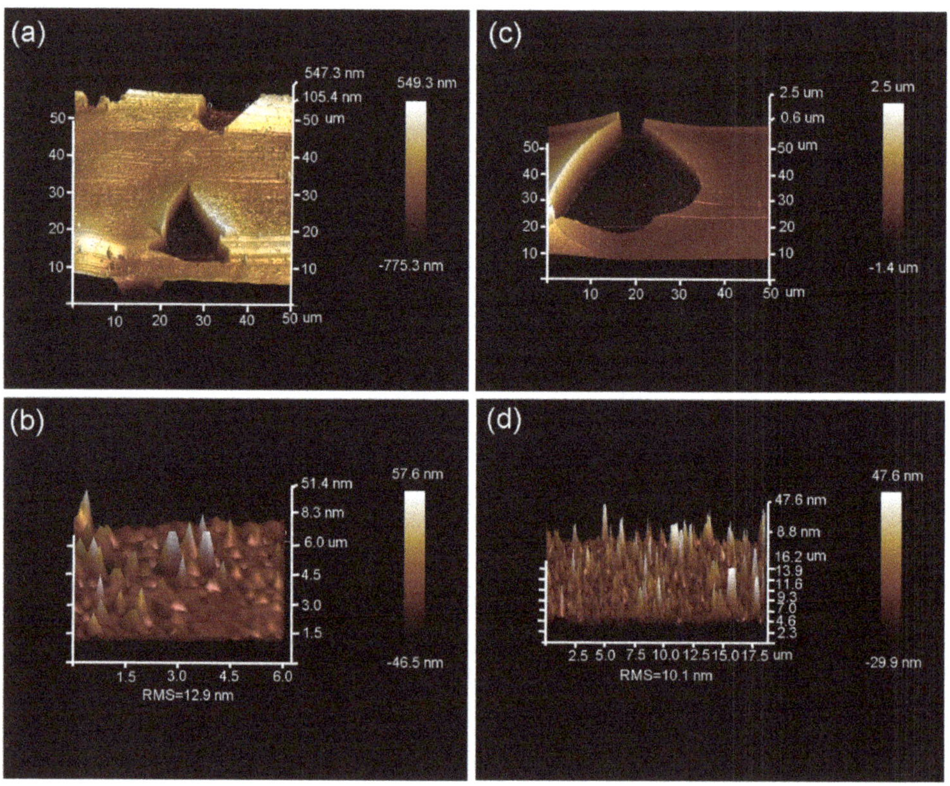

Figure 4. AFM images showing over-viewed surface morphology with a scanning zone of 50 × 50 μm² of the 194 keV H_2^+-implanted 6H-SiC at 450 °C followed by 1100 °C annealing for 15 min (**a**), the 194 keV H_2^+-implanted 6H-SiC at 900 °C (**c**); (**b**,**d**) show the surface morphology of the exfoliation zone for the H_2^+-implanted 6H-SiC at 450 °C followed by 1100 °C annealing for 15 min and H_2^+-implanted 6H-SiC at 900 °C, respectively.

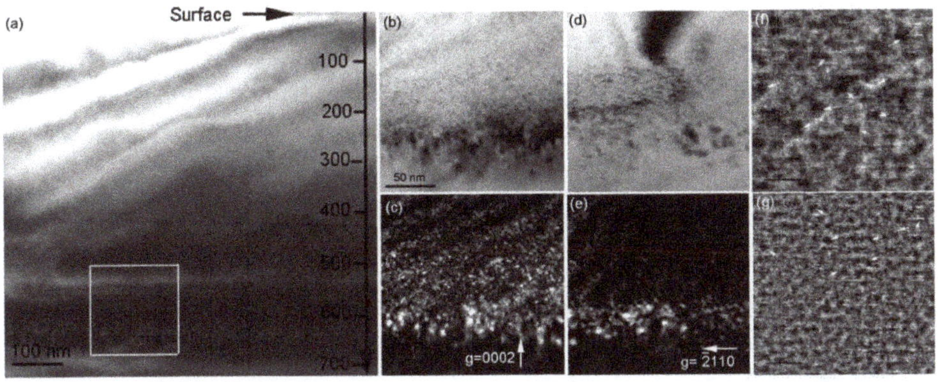

Figure 5. Bright-field XTEM image showing the over-viewed damage distribution in the 194 keV $H_2{}^+$-implanted 6H-SiC at 450 °C. The enlarged selected zone shown in (**a**) is presented in (**b**) to (**e**), where (**b**) BF and (**c**) WBDF images were observed under g = 0002 and (**d**,**e**) were observed under $g = \bar{2}110$. (**f**) The high-resolution transmission electron microscope image shows lattice defects exhibiting a dark contrast, as indicated by white arrows and (**g**) inverse Fourier-filtered image of (**f**) in order to improve the visibility of lattice defects. (**c**–**e**) have the same scale as (**b**).

Figure 6 presents the damage distribution in the 194 keV $H_2{}^+$-implanted 6H-SiC at 450 °C after 1100 °C annealing for 15 min. Compared with the microstructure observed in the as-implanted sample, after 1100 °C annealing, three evident changes can be observed. The first change is the depth distribution of the damage band. Figure 6a shows the over-viewed damage distribution where the damage band located at a depth ranging from 510 to 590 nm beneath the sample surface can be clearly distinguished. The observed damage band is shallower than that of the $H_2{}^+$-implanted 6H-SiC at 450 °C. This is attributed to defect recovery after 1100 °C annealing—Frank loops in particular. The observed damage band is well consistent with the simulated profiles. The second change is that a long microcrack parallel to the sample surface is observed in the front of the damage band. The formation of microcracks is due to the combined effects of Si vacancy migration at 1100 °C and the chemical interaction of H atoms and dangling bonds in the platelets [31,32]. The third change is that the width of the damage band observed under g = 0002 is the same as the case under $g = \bar{2}110$. The width of the damage band observed under $g = \bar{2}110$ increases after annealing. This is a reverse annealing phenomenon that is attributed to the growth of the microcrack accompanied by emitting interstitial atoms. A similar phenomenon was observed in He-implanted SiC [33,34]. The HRTEM image shown in Figure 6f confirms an amorphous structure inside the microcrack. Interstitial-type dislocation loops formed in the periphery of the microcrack are shown in Figure 6g. The microcrack is not straight, which induces the roughness of the exfoliation surface. The exfoliation surface was measured by AFM, and the result is shown in Figure 4.

Figure 7 presents the damage distribution in the 194 keV $H_2{}^+$-implanted 6H-SiC at 900 °C. It can be seen in Figure 7a that a damage band is located at a depth ranging from 400 to 620 nm beneath the surface. Inside the damage band, a microcrack exhibiting bright contrast is located at 558 nm beneath the surface. Around the microcrack, dense Frank loops were observed, as shown in Figure 7d–g. The HRTEM image shows an amorphous structure inside the microcrack. Compared with the case of $H_2{}^+$-implanted 6H-SiC at 450 °C followed by 1100 °C annealing for 15 min, there are two significant differences in the 6H-SiC implanted with $H_2{}^+$ ions at 900 °C. The first difference is the location of the microcrack. The microcrack is in the front of the damage band of the 6H-SiC implanted with $H_2{}^+$ ions at 450 °C followed by 1100 °C annealing, whereas the microcrack is located at a depth between the damage peak and maximum hydrogen deposition simulated by SRIM-2013 for the 6H-SiC implanted with $H_2{}^+$ ions at 900 °C. The second difference is the width of the damage band. The width of the damage band increases significantly during implantation at 900 °C. Two reasons can account for this. One is the increasing vacancy-hydrogen interaction at 900 °C, but not at 450 °C [31]. The growth of vacancy-hydrogen clusters can emit interstitials and then these interstitials migrate and accumulate into Frank loops. The fast growth of the vacancy-hydrogen clusters leads to the occurrence of the microcrack, as observed in Figure 7a. The other is the influence of the sample surface, which acts as a defect sink. As shown in Figure 7a, some lattice defects were observed at a depth near 400 nm. It is indicated that some interstitials produced by H_2^+ collision migrate towards the sample surface during implantation at 900 °C. This is consistent with the defect distribution in the He-implanted SiC at elevated temperatures [35].

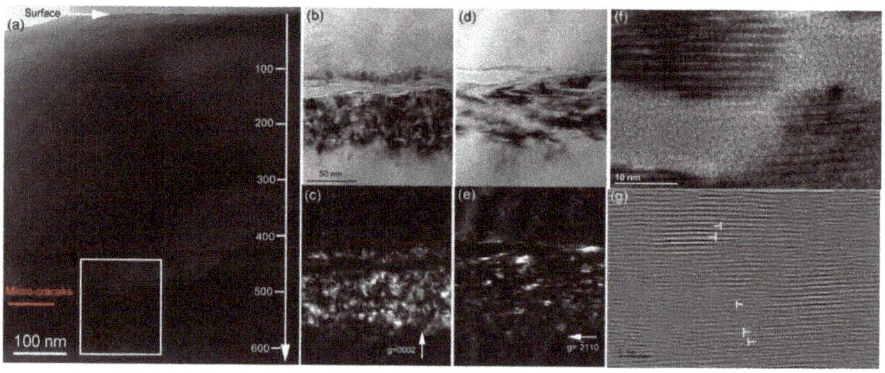

Figure 6. (**a**) Bright-field XTEM image showing the over-viewed damage distribution in the 194 keV H_2^+-implanted 6H-SiC at 450 °C followed by 1100 °C annealing for 15 min. The enlarged selected zone shown in (**a**) is presented in (**b**–**e**), where (**b**) BF and (**c**) WBDF images were observed under g = 0002, and (**d**,**e**) images were observed under $g = \bar{2}110$. (**f**) The high-resolution transmission electron microscope image shows the lattice fringe along the micro-crack, and (**g**) the inverse Fourier-filtered image of (**f**) shows interstitial-type dislocation loops on the (0001) plane. The scale in (**b**) is the same as that in (**c**–**e**).

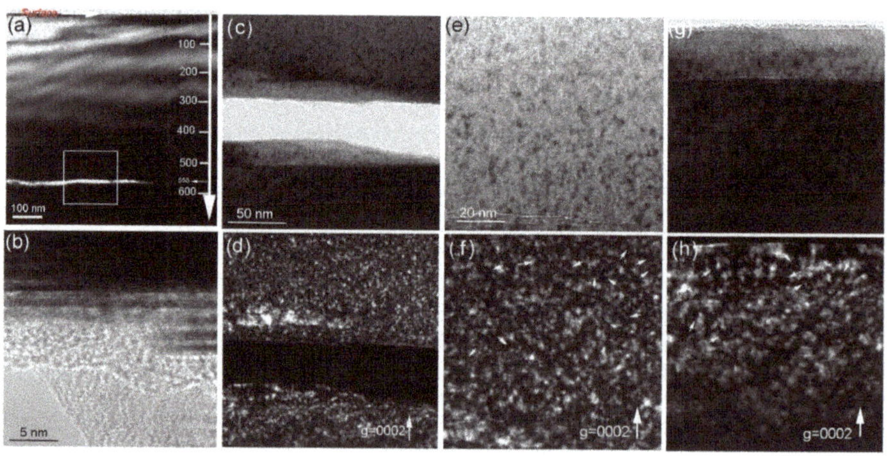

Figure 7. (**a**) Bright-field XTEM image showing the over-viewed damage distribution in the 194 keV H_2^+-implanted 6H-SiC at 900 °C. (**c**) BF image and (**d**) WBDF image under g = 0002 show the microstructure in the periphery of the microcrack. The magnified images taken from (**a**) show the lattice defects indicated by arrows above the microcrack presented in (**e**,**f**) and below the microcrack presented in (**g**,**h**). (**b**) HRTEM image shows the lattice fringe along the microcrack.

4. Discussion

Surface blisters and exfoliation of hydrogen implantation into 6H-SiC are attributed to the growth of microcracks inside the sample. When the amount of hydrogen is sufficient, this leads to internal pressure that is high enough to overcome the surface energy γ, and then to open up the crack. Matani and Gosele [36] have argued the critical radius for the on-set of blistering.

$$r_{cric} = \left\{ \frac{16\gamma E t^3}{9\alpha(1-v^2)\Delta p^2} \right\}^{1/4} \tag{1}$$

where Δp is the difference between the inside platelets and the outside atmosphere, t is the microcrack depth, E is the material's Young's modulus, v is Poisson's ratio, α is a numerical factor in the order of ~1, and γ is the specific interface energy which would be changed by the implantation-induced defects. Freund [37] developed a model to explain wafer splitting via crack growth triggered by gas pressure p. The necessary condition for crack growth can be expressed as:

$$p = \mu\left(\frac{\pi}{1-v}\frac{\gamma}{a\mu}\right)^{1/2} \tag{2}$$

where μ is the shear modulus, and a is the crack size. Based on Equation (2), the critical pressure required in the crack cavity decreases with the increase in the crack size.

The microstructure shows the size of the observed microcrack is larger in the 6H-SiC implanted with H_2^+ ions at 900 °C than that in the H_2^+-implanted 6H-SiC at 450 °C followed by 1100 °C annealing. This result indicates the fast growth of microcracks when the sample was implanted at a higher temperature, compared with the lower temperature implantation followed by a higher temperature annealing. According to Equation (2), the critical value of the inner pressure was easily achieved when the sample was implanted at 900 °C; therefore, the critical radius for the on-set of blistering increased when the sample was implanted at 900 °C based on Equation (1). Because the size of an exfoliation zone is smaller than its corresponding blister, the observed exfoliation size was far larger when the sample was implanted at 900 °C, as compared to the sample implanted at 450 °C and then annealed at 1100 °C.

Unlike C and Si interstitials, C and Si vacancies cannot migrate at 450 °C [31]. Dynamic annealing is, therefore, not significant and many survival defects are formed inside the sample. These defects lead to the evident lattice swelling. After annealing at 1100 °C for 15 min, an obvious microcrack was observed in the front of the damage band, where the lattice damage was smaller than the damage peak. It can be speculated that the specific interface energy γ increases with the increase in lattice defects [38]; therefore, the increase in γ needs more gas pressure of the crack in order to sustain crack growth. As a result, the crack growth is retarded in the peak damage region. When the sample was implanted at 900 °C, interstitials and vacancies can migrate simultaneously. The compound of hydrogen and vacancies can rapidly form and then coalesce into platelets. The growth of platelets is followed by pushing away of the matrix atoms. Named after the trap-mutation process [39,40], this can be expressed as:

$$H_nV_m \rightarrow H_nV_{m+1} + I \tag{3}$$

Due to the rapid growth of platelets, dense self-interstitials were pushed away to form dislocation loops, which were observed by XTEM, as shown in Figure 7. This implies that the damage band is wider and the concentration of the observed lattice defects is higher in the sample implanted at 900 °C compared to the sample implanted at 450 °C. It should be noted that lattice swelling was significant when the sample was implanted at 450 °C, but not 900 °C. This is attributed to the limited resolution of conventional TEM (near 1 nm in WBDF), and hence there are many formed interstitials due to slow growth at 450 °C, which are too small to be observed by XTEM. After 1100 °C annealing, some interstitials migrated and recombined with vacancies to recover the lattice damage. Other interstitials migrated and coalesced into dislocation loops, which were observed under WBDF with $g = \bar{2}110$, as shown in Figure 6e. It should be discussed why the microcrack is located at a depth between the damage peak and the maximum hydrogen deposition for the sample implanted at 900 °C. Because of the rapid growth of vacancy-hydrogen clusters at 900 °C, available vacancies and hydrogen were higher in the middle of the two peaks than in other zones, and thus a microcrack easily grew in this zone, which was observed by XTEM (see Figure 7a).

The scanning white-light interferometry method showed that the exfoliation depth in the different exfoliation sites was almost the same. The formation of hillocks after exfoliation was due to the crack growth via the coalescence of platelets in the different depths; therefore, a zigzagged microcrack was observed by HRTEM, as shown in Figures 6f and 7h. In the microcrack zone, an amorphous structure demonstrated the strong interaction between hydrogen and carbon/silicon dangling bonds on the inner surface of the microcrack. This is consistent with the report of Hojou et al. [41] that H_2, C-H, and Si-H compounds were formed in the bubbles produced by H_2^+ + He^+ simultaneously-implanted polycrystalline 6H-SiC. Therefore, it can be argued that hydrogen is effective in enhancing amorphization due to the chemical interaction. The surface morphology showed the exfoliation efficiency was higher for the sample implanted at 900 °C than that of the sample implanted 450 °C and subsequently annealed at 1100 °C. However, the formed damage band in the sample implanted at 900 °C was almost twice as high as that in the sample implanted at 450 °C and subsequently annealed at 1100 °C. The formed dislocation loops are stable and not easily annealed after thermal treatment [42–44]. To fabricate the SiCOI structure for the final electronic and optoelectronic device applications, these lattice damage zones after wafer transfer must be removed by the chemical mechanical polishing (CMP) process. To enable reuse of the wafer, the survival damage band should as narrow as possible. Therefore, low-temperature implantation, followed by high-temperature annealing is a better choice for the fabrication of the SiCOI structure compared with conventional implantation at a high temperature.

It should be noted that in nuclear fusion applications, dense energetic hydrogen can be produced in SiC by nuclear transmutations [45]. This hydrogen can interact with SiC forming displacement damage cascades and subsequently deposits in the near-surface layer. Because the first wall of the fusion reactor is expected to face a very high temperature, a comprehensive understanding of the surface exfoliation of SiC will require further study involving irradiation experiments at different temperatures and utilizing ion fluxes.

5. Conclusions

The exfoliation efficiency of H_2^+ implantation at 450 °C and 900 °C to a fluence of 5×10^{16} H_2/cm^2 in 6H-SiC was investigated. A lattice swelling of 5% was observed in the H_2^+-implanted 6H-SiC at 450 °C. Reverse dynamic annealing was observed in the H_2^+-implanted 6H-SiC at 900 °C. This is related to the rapid growth of hydrogen and vacancy clusters (H_nV_m), following by emitting interstitials around H_nV_m. A microcrack was observed in the front of the damage band in the sample implanted with H_2^+ ions at 450 °C and subsequently annealed at 1100 °C for 15 min. In the sample implanted with H_2^+ ions at 900 °C, a microcrack occurred between the displacement damage peak and the maximum hydrogen deposition. The change in specific interface energy γ can explain the location of the microcrack. Despite a high efficiency of exfoliation in the sample implanted at 900 °C, this procedure is not considered optimal for the fabrication of the SiCOI structure due to a wide damage band formed during H_2^+ implantation. Instead, we propose hydrogen implantation at a temperature lower than the critical temperature for vacancy migration as a more suitable method for this purpose.

Author Contributions: Data curation and funding acquisition, T.W.; data curation and concepturalization, Z.Y.; original writing & editing, supervision and project administration, B.L.; data curation, S.X. and Q.L.; formal analysis, F.G.; data curation and resources, T.Z. and J.L. All authors have read and agreed to the published version of the manuscript.

Funding: The work was funded by the National Natural Science Foundation of China (Grant Nos. U1832133, 11905206), Sichuan Science and Technology Program (Grant No. 2020ZYD055), and the Doctor Research Foundation of Southwest University of Science and Technology (Grant No. 18zx714101).

Conflicts of Interest: The authors declare no conflict of interest.

References

1. Wesch, W. Silicon carbide: Synthesis and processing. *Nucl. Instrum. Methods Phys. Res. B* **1996**, *116*, 305–321. [CrossRef]
2. Raynaud, C. Silica films on silicon carbide: A review of electrical properties and device applications. *J. Non-Cryst. Solids* **2001**, *280*, 1–31. [CrossRef]
3. Celler, G.K.; Cristoloveanu, S. Frontiers of silicon-on-insulator. *J. Appl. Phys.* **2003**, *93*, 4955–4978. [CrossRef]
4. Bruel, M. Application of hydrogen ion beams to silicon on insulator material technology. *Nucl. Instrum. Methods Phys. Res. B* **1996**, *108*, 313–319. [CrossRef]
5. Herley, R.E.; Suder, S.; Gamble, H.S. Ion implantation of hydrogen and helium into silicon wafers for layer transfer in devices. *Vacuum* **2005**, *78*, 167–175. [CrossRef]
6. Taraschi, G.; Pitera, A.J.; Fitzgerald, E.A. Strained Si, SiGe, and Ge on-insulator: Review of wafer bonding fabrication techniques. *Solid-State Electron.* **2004**, *48*, 1297–1305. [CrossRef]
7. Woo, H.J.; Choi, H.W.; Kim, J.K.; Kim, G.D.; Hong, W.; Choi, W.B.; Bae, Y.H. Thick Si-on-insulator wafers formation by ion-cut process. *Nucl. Instrum. Methods Phys. Res. B* **2005**, *241*, 531–535. [CrossRef]
8. Tong, Q.Y.; Gutjahr, K.; Hopfe, S.; Gosele, U.; Lee, T.H. Layer splitting process in hydrogen-implanted Si, Ge, SiC, and diamond substrates. *Appl. Phys. Lett.* **1997**, *70*, 1390–1392. [CrossRef]
9. Zhang, L.; Li, B.S. Study of surface exfoliation on 6H-SiC induced by H_2^+ implantation. *Physica B* **2017**, *508*, 104–111. [CrossRef]
10. Gregory, R.B.; Wetteroth, T.A.; Wilson, S.R.; Holland, O.W.; Thomas, D.K. Effects of irradiation temperature and dose on exfoliation of H^+-implanted silicon carbide. *Appl. Phys. Lett.* **1999**, *75*, 2623–2625. [CrossRef]
11. Bennett, J.A.; Holland, O.W.; Budde, M.; Thomas, D.K.; Feldman, L.C. Complete surface exfoliation of 4H-SiC by H^+-and Si^+-coimplantation. *Appl. Phys. Lett.* **2000**, *76*, 3265–3267. [CrossRef]
12. Amarasinghe, V.P.; Wielunski, L.; Barcz, A.; Feldman, L.C.; Celler, G.K. Optimization of H^+ implantation parameters for exfoliation of 4H-SiC films. *ECS Trans.* **2012**, *50*, 341–348. [CrossRef]
13. Daghbouj, N.; Li, B.S.; Callisti, M.; Sen, H.S.; Lin, J.; Ou, X.; Karlik, M.; Polcar, T. The structural evolution of light-ion implanted 6H-SiC single crystal: Comparison of the effect of helium and hydrogen. *Acta Mater.* **2020**, *188*, 609–622. [CrossRef]
14. Daghbouj, N.; Li, B.S.; Karlik, M.; Declemy, A. 6H-SiC blistering efficiency as a function of the hydrogen implantation fluence. *Appl. Surf. Sci.* **2019**, *466*, 141–150. [CrossRef]
15. Liu, H.P.; Li, J.Y.; Li, B.S. Microstructure of hydrogen-implanted polycrystalline a-SiC after annealing. *Chin. Phys. Lett.* **2018**, *35*, 096103–096105. [CrossRef]
16. Senga, K.; Kimoto, T.; Suda, J. Hydrogen implantation and annealing-induced exfoliation process in SiC wafers with various crystal orientations. *Jpn. J. App. Phys.* **2008**, *47*, 5352–5354. [CrossRef]
17. Jia, Q.; Huang, K.; You, T.G.; Yi, A.L.; Lin, J.J.; Zhang, S.B.; Zhou, M.; Zhang, B.; Yu, W.J.; Ou, X.; et al. Freestanding ultrathin single-crystalline SiC substrate by MeV H ion-slicing. *Appl. Phys. Lett.* **2018**, *112*, 192102–192104. [CrossRef]
18. Interactions of Ions with Matter. Available online: http://www.srim.org (accessed on 7 March 2013).
19. Nakashima, S.; Harima, H. Raman investigation of SiC polytypes. *Phys. Stat. Sol. (a)* **1997**, *162*, 39–64. [CrossRef]
20. Sorieul, S.; Costantini, J.M.; Gosmain, L.; Thome, L.; Grob, J.J. Raman spectroscopy study of heavy-ion-irradiated α-SiC. *J. Phys. Condens. Matter* **2006**, *18*, 5235–5251. [CrossRef]
21. Li, B.S.; Zhang, C.H.; Zhang, H.H.; Shibayama, T.; Yang, Y.T. Study of the damage produced in 6H-SiC by He irradiation. *Vacuum* **2011**, *86*, 452–456. [CrossRef]
22. Li, B.S.; Wang, Z.G. Structures and optical properties of H_2^+-implanted GaN epi-layers. *J. Phys. D Appl. Phys.* **2015**, *48*, 225101–225112. [CrossRef]

23. Richter, H.; Wang, Z.P.; Ley, L. The one phonon Raman spectrum in microcrystalline silicon. *Solid State Commun.* **1981**, *39*, 625–629. [CrossRef]
24. Parayanthal, P.; Pollak, F.H. Raman scattering in alloy semiconductors: "Spatial Correlation" model. *Phys. Rev. Lett.* **1984**, *52*, 1822–1825. [CrossRef]
25. Koyanagi, T.; Lance, M.J.; Katoh, Y. Quantification of irradiation defects in beta-silicon carbide using Raman spectroscopy. *Scr. Mater.* **2016**, *125*, 58–62. [CrossRef]
26. Dadwal, U.; Singh, R. On the mechanism of blistering phenomenon in high temperature H-implanted GaN. *Appl. Phys. Lett.* **2013**, *102*, 081606–081610. [CrossRef]
27. Dadwal, U.; Kumar, P.; Moutanabbir, O.; Reiche, M.; Singh, R. Effect of implantation temperature on the H-induced microstructural damage in AlN. *J. Alloy. Compd.* **2014**, *588*, 300–304. [CrossRef]
28. Li, B.S.; Zhang, C.H.; Zhang, H.H.; Zhang, Y.; Yang, Y.T.; Zhou, L.H.; Zhang, L.Q. Thermodynamic model of helium and hydrogen co-implanted silicon surface layer splitting. *Nucl. Instrum. Methods Phys. Res. B* **2010**, *268*, 555–559. [CrossRef]
29. Ishimaru, M.; Bae, I.-T.; Hirata, A.; Hirotsu, Y.; Valdez, J.A.; Sickafus, K.E. Volume swelling of amorphous SiC during ion-beam irradiation. *Phys. Rev. B* **2005**, *72*, 024116–024122. [CrossRef]
30. Leclerc, S.; Declemy, A.; Beaufort, M.F.; Tromas, C.; Barbot, J.F. Swelling of SiC under helium implantation. *J. Appl. Phys.* **2005**, *98*, 113506–113511. [CrossRef]
31. Gao, F.; Weber, W.J.; Posselt, M.; Belko, V. Atomistic study of intrinsic defect migration in 3C-SiC. *Phys. Rev. B* **2004**, *69*, 245205–245209. [CrossRef]
32. Muto, S.; Tanabe, T.; Maruyama, T. Cross sectional TEM observation of gas-ion-irradiation induced surface blisters and their precursors in SiC. *Mater. Trans.* **2003**, *44*, 2599–2604. [CrossRef]
33. Daghbouj, N.; Li, B.S.; Callisti, M.; Sen, H.S.; Karlik, M.; Polcar, T. Microstructural evolution of helium-irradiated 6H-SiC subjected to different irradiation conditions and annealing temperatures: A multiple characterization study. *Acta Mater.* **2019**, *181*, 160–172. [CrossRef]
34. Li, B.S.; Du, Y.Y.; Wang, Z.G.; Wei, K.F.; Zhang, H.P.; Yao, C.F.; Chang, H.L.; Sun, J.R.; Cui, M.H.; Sheng, Y.B.; et al. Transmission electron microscopy investigations of bubble formation in He-implanted polycrystalline SiC. *Vacuum* **2015**, *113*, 75–83. [CrossRef]
35. Li, B.S.; Liu, H.P.; Shen, T.L.; Xu, L.J.; Wang, J.; Zhao, F.W.; Peng, D.P.; Li, J.H.; Sheng, Y.B.; Xiong, A.L. Irradiation-induced microstructure damage in He-irradiated 3C-SiC at 1000 °C. *J. Eur. Ceram. Soc.* **2020**, *40*, 1014–1022. [CrossRef]
36. Mitani, K.; Gosele, U.M. Formation of interface bubbles in bonded silicon wafers: A thermodynamic model. *Appl. Phys. A* **1992**, *54*, 543–552. [CrossRef]
37. Freund, L.B. A low bound on implant density to induce wafer splitting in forming compliant substrate structures. *Appl. Phys. Lett.* **1997**, *70*, 3519–3521. [CrossRef]
38. Hochbauer, T.; Mirsa, A.; Nastasi, M.; Mayer, J.W. Physical mechanisms behind the ion-cut in hydrogen implanted silicon. *J. Appl. Phys.* **2002**, *92*, 2335–2342. [CrossRef]
39. Hartmann, M.; Trinkaus, H. Evolution of gas-filled nanocracks in crystalline solids. *Phys. Rev. Lett.* **2002**, *88*, 055505–055508. [CrossRef]
40. Oliviero, E.; Beaufort, M.F.; Barbot, J.F. Influence of dose rate on bubble formation by high energy He implantation in silicon. *J. Appl. Phys.* **2001**, *90*, 1718–1724. [CrossRef]
41. Hojou, K.; Furuno, S.; Kushita, K.N.; Sasajima, N.; Izui, K. EELS analysis of SiC crystals under hydrogen and helium dual-ion beam irradiation. *Nucl. Instrum. Method Phys. Res. B* **1998**, *141*, 148–153. [CrossRef]
42. Sawabe, T.; Akiyoshi, M.; Ichikawa, K.; Yoshida, K.; Yano, T. Microstructure of heavily neutron-irradiated SiC after annealing up to 1500 °C. *J. Nucl. Mater.* **2009**, *386–388*, 333–337. [CrossRef]
43. Ye, C.; Ran, G.; Zhou, W.; Shen, Q.; Feng, Q.J.; Lin, J.X. Recrystallization-induced surface cracks of carbon ions irradiated 6H-SiC after annealing. *Materials* **2017**, *10*, 1231. [CrossRef] [PubMed]
44. Chen, J.; Jung, P.; Klein, H. Production and recovery of defects in SiC after irradiation and deformation. *J. Nucl. Mater.* **1998**, *258–263*, 1803–1808. [CrossRef]

45. Sawan, M.E.; Ghoniem, N.M.; Snead, L.; Katoh, Y. Damage production and accumulation in SiC structures in inertial and magnetic fusion systems. *J. Nucl. Mater.* **2011**, *417*, 445–450. [CrossRef]

Publisher's Note: MDPI stays neutral with regard to jurisdictional claims in published maps and institutional affiliations.

materials MDPI

Article

Application of Proton Irradiation in the Study of Accelerated Radiation Ageing in a GaAs Semiconductor

Igor Neuhold [1,2,*], Pavol Noga [3], Stanislav Sojak [1,3], Martin Petriska [1], Jarmila Degmova [1,3], Vladimir Slugen [1] and Vladimir Krsjak [1,3,*]

[1] Institute of Nuclear and Physical Engineering, Faculty of Electrical Engineering and Information Technology, Slovak University of Technology in Bratislava, Ilkovicova 3, 81219 Bratislava, Slovakia
[2] European Organization for Nuclear Research (CERN), 1211 Geneva, Switzerland
[3] Advanced Technologies Research Institute, Faculty of Materials Science and Technology, Slovak University of Technology in Bratislava, Jana Bottu 25, 91724 Trnava, Slovakia
* Correspondence: igor.neuhold@stuba.sk (I.N.); vladimir.krsjak@stuba.sk (V.K.)

Abstract: Proton irradiation experiments have been used as a surrogate for studying radiation effects in numerous materials for decades. The abundance and accessibility of proton accelerators make this approach convenient for conducting accelerated radiation ageing studies. However, developing new materials with improved radiation stability requires numerous model materials, test samples, and very effective utilization of the accelerator beam time. Therefore, the question of optimal beam current, or particle flux, is critical and needs to be adequately understood. In this work, we used 5 MeV protons to introduce displacement damage in gallium arsenide samples using a wide range of flux values. Positron annihilation lifetime spectroscopy was used to quantitatively assess the concentration of radiation-induced survived vacancies. The results show that proton fluxes in range between 10^{11} and 10^{12} $cm^{-2}.s^{-1}$ lead to a similar concentration of monovacancies generated in the GaAs semiconductor material, while a further increase in the flux leads to a sharp drop in this concentration.

Keywords: semiconductors; WBG; proton irradiation; ageing; gallium arsenide; positron annihilation spectroscopy

Citation: Neuhold, I.; Noga, P.; Sojak, S.; Petriska, M.; Degmova, J.; Slugen, V.; Krsjak, V. Application of Proton Irradiation in the Study of Accelerated Radiation Ageing in a GaAs Semiconductor. *Materials* **2023**, *10*, 1089. https://doi.org/10.3390/ma16031089

Academic Editor: Scott M. Thompson

Received: 21 December 2022
Revised: 17 January 2023
Accepted: 21 January 2023
Published: 27 January 2023

1. Introduction

In the last several decades, dominated by silicon (Si) and gallium arsenide (GaAs), semiconductors have shaped the new technological era with diodes, transistors, and integrated circuits [1]. Gradually, semiconductor technology has entered all industry areas, including nuclear power production. While the previous generation of nuclear power plants restricted the use of electronic devices to an inevitable minimum, recent nuclear plants rely on the electronics used not just in the digital computers and process control systems in a mild environment, but also in harsh radiation conditions, whereas the use of different electronic systems is not limited to detector technology only.

The application of semiconductors in harsh radiation environments is significantly increasing, not just by nuclear power plants, but also in medical diagnostics, nuclear science, technology, research, and space applications. In all of these fields, high-energy charged particles interact with essential safety and other components, modifying their microstructure and affecting their lifetime. Therefore, the need for safe long-term operation of the semiconductors is crucial for the reliability of electronic instruments, and any failure in critical components leads to substantial economic and human safety hazards in all of these applications.

Despite their susceptibility to permanent degradation and catastrophic failure due to heavy-ion exposure [2], numerous research publications have already pointed out that future semiconductor technologies, including those for space, detectors, medicine, and nuclear applications, consider the application of wide band gap (WBG) semiconductors such as GaN and SiC. In these crystals, the gap between the valence and conduction bands is an essential parameter that defines not only the electrical properties, but also the susceptibility to radiation [3]. The advantage of WBG compared with classical semiconductors such as silicon and gallium arsenide is in the improved electrical properties, such as a higher efficiency, switching frequency, operating temperature, and higher operating voltage [4,5]. This leads to faster, dimension-wise, smaller, more powerful, and more efficient components. These capabilities will be reflected in smaller sizes and weights and will have less power demand due to limited power losses [3,6].

While natural radiation environments, such as the ionosphere, trapped radiation belts, solar particle events, and galactic cosmic rays dominate in outer space, on the ground, various man-made applications lead to the exposure of semiconductor materials to ionizing radiation. Although the understanding of their radiation tolerance is far from complete, silicon carbide (SiC) and gallium nitride (GaN) semiconductors are expected to have superior electrical properties, and their susceptibility to harsh radiation environments compared with the more conventional semiconductors needs to be understood in more detail. It could be expected that because of the improved electrical and radiation properties, GaN has excellent potential to improve a safe long-term operation, decreasing the life-cycle cost and lowering the occurrence of failure, which could lead to personal safety risks. [7,8].

Radiation effects in semiconductor-based electronics due to harsh radiation environments can be divided into two categories, namely the short-term temporary effects and the long-term permanent degradation. The short-term temporary effect comes mainly from the effects of ionising energy loss in the semiconductor by the energetic particle, causing single event effects (SEEs) such as single event upset (SEU), single event transient (SET), single event latch-up (SEL), single event gate rupture (SEGR), or single event burnout (SEB). On the other hand, the long-term effects are dominantly created by the non-ionising energy losses in the material by the displacement damage, where elastic collisions with the material can eject atoms from their standard position in the lattice or when primary recoil atoms collide with other atoms in the lattice [9,10].

While "realistic" low-dose long-term irradiation experiments provide reliable data for assessing the electronic components and circuits resistant to damage or malfunction caused by high levels of ionizing radiation, their potential is significantly decreased for fast technological development due to the time-consuming nature of the approach, which also represents a significant cost in the radiation experiment. To guarantee reliable long-term operation in harsh radiation environments for a reasonable duration of the experiment, semiconductors must undergo suitable accelerated ageing tests. A proper accelerating radiation ageing mechanism is necessary among other ageing mechanisms such as thermal and mechanical vibration, contributing to the successful assessment of the lifetime of electronic devices. However, there has not been an engineering consensus yet on how the results of accelerated ageing experiments can be extrapolated to the engineering and design of technologies for long-term applications. A deep understanding of the evolution of the microstructure exposed to accelerated radiation tests inevitably requires employing both theoretical modelling and suitable experimental characterisation methods sensitive to the atomic-scale lattice defects. This is a very complicated and challenging task due to the limited size sensitivity of the experimental techniques on the one hand, and the limited size of the theoretical calculation models on the other.

For the characterisation of the material damage, a positron annihilation spectroscopy using a ^{22}Na positron source was used. Positron annihilation spectroscopy (PAS) has been used as a microstructural characterisation tool that is sensitive to vacancy-type defects. This technique has been widely used in characterising various types of defects in semiconductors since the 1970s. Positron annihilation experiments were successfully used in the characterisation of radiation effects in (not only) semiconductors modified in numerous types of radiation experiments [11,12], including gamma radiation [13], electron irradiation, and neutron irradiation [14], as well as proton irradiation [15]. In this paper, we used this technique to obtain a quantitative characterisation of the radiation-induced vacancy-type defects, which were investigated as a function of the proton flux (displacement damage rate).

This work aims to explore the feasibility of using proton implantation as a mechanism for the radiation ageing of semiconductors and to improve the understanding of the process of creating displacement damage in bulk GaAs semiconductor material. The particular goal of the study is to describe the role of the "flux effect" on the evolution of the microstructure. In other words, the work was aimed at achieving a better understanding of how to optimise accelerated-ageing irradiation experiments in order to make them a physically meaningful representation of the long-term permanent degradation of the material exposed in the radiation field, so as to establish a comparison between the produced defects and surviving defects in the irradiated materials. The "flux effect" on the WBG semiconductors will be investigated in our forthcoming study, and will be compared with present work.

2. Experimental

2.1. Material and Sample Preparation

The GaAs samples investigated in this study were cut into dimensions of ~10 × 10 mm, from a monocrystalline wafer obtained from a local manufacturer—The Gallium Arsenide Company Slovakia, CMK Ltd. (Zarnovica, Slovakia). Detailed information on the material, provided by the manufacturer, is shown in Table 1.

For the irradiation experiment and subsequent PAS characterisation, in a total, six pairs of samples were required, while one sample pair served as a reference. Two samples were destroyed during the cutting of the sample (Figure 1), and some samples were damaged and thus could not be used for the study evaluated during the proton irradiation.

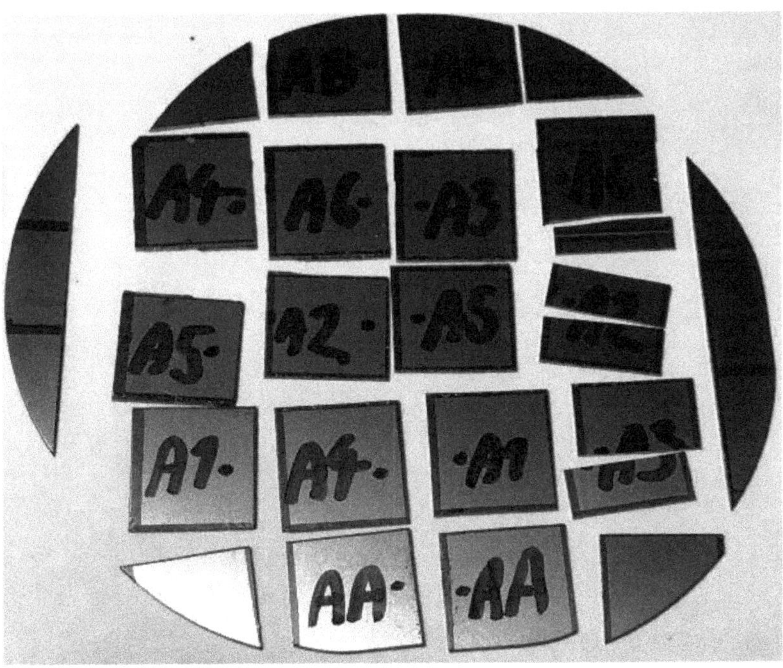

Figure 1. As-prepared test samples of the GaAs as per Table 1 for the irradiation experiment.

Table 1. Specification of the studied GaAs samples.

Parameter	Properties
Product No. Method Description Type	82045 GaAs LEC Monocrystalline wafers Semi-insulating, undoped
Dopant	N/A
Resistivity Hall Mobility	4.38×10^8 Ohm cm 5398 $Cm^2\ V^{-1}$ s
Orientation Off Orientation	$(100) \pm 0.5°$ Off 2° towards (110)°
Diameter Thickness	50.8 ± 0.1 mm 500 ± 25 μm
Surface Front side Back side	SSP Polished Lapped/etched

2.2. *Proton Irradiation Experiment*

The proton irradiation experiment was performed using the 6 MV Tandetron tandem accelerator at the STU University Science Park CAMBO located in Trnava (Figure 2). The accelerator is used for a wide range of ion irradiation studies including H, He, and heavy ion irradiation. The maximum achievable energy for proton irradiation is 12 MeV and the maximum flux, depending on beam scanning area, which can reach up to 10^{14} $cm^{-2}.s^{-1}$ [16].

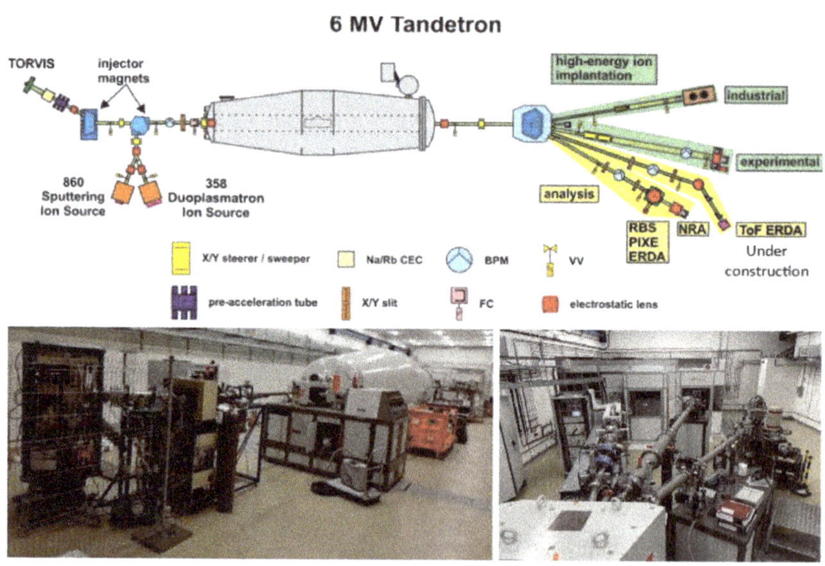

Figure 2. The 6 MV Tandetron ion accelerator at the Advanced Technologies Research Institute, Slovak University of Technology [17].

The actual irradiation times and corresponding proton fluxes shown in Table 2 were proposed according to the availability of the accelerator, in order to obtain a wide range of proton fluxes that will be increasing logarithmically. While different target fluences were initially considered for this experiment, finally a fluence of 10^{16} cm^{-2} was selected to be achieved by using five different fluxes (ranging from 10^{11} to 11^{13} $cm^{-2}.s^{-1}$). The energy of the protons in Figure 3 is discussed later in this chapter. The reasoning behind the selection of the fluence is based on the sensitivity of the PALS technique and it is illustrated in Figure 4.

Table 2. As-prepared test samples for the irradiation experiment.

Sample Set No.	Irradiation Time [min]/[h]	Flux [$cm^{-2}.s^{-1}$]
Set 0 (reference)	0/0	0
Set 1	16/0.27	1.04×10^{13}
Set 2	36/0.59	4.63×10^{12}
Set 3	222/3.70	7.51×10^{11}
Set 4	505/8.42	3.30×10^{11}
Set 5	1599/26.65	1.04×10^{11}

Note, that in two cases (sets 2 and 4), the samples measured using the PAS technique were not identical in terms of the proton fluence received. The proton flux for these samples was calculated as the mean of the two nearest values.

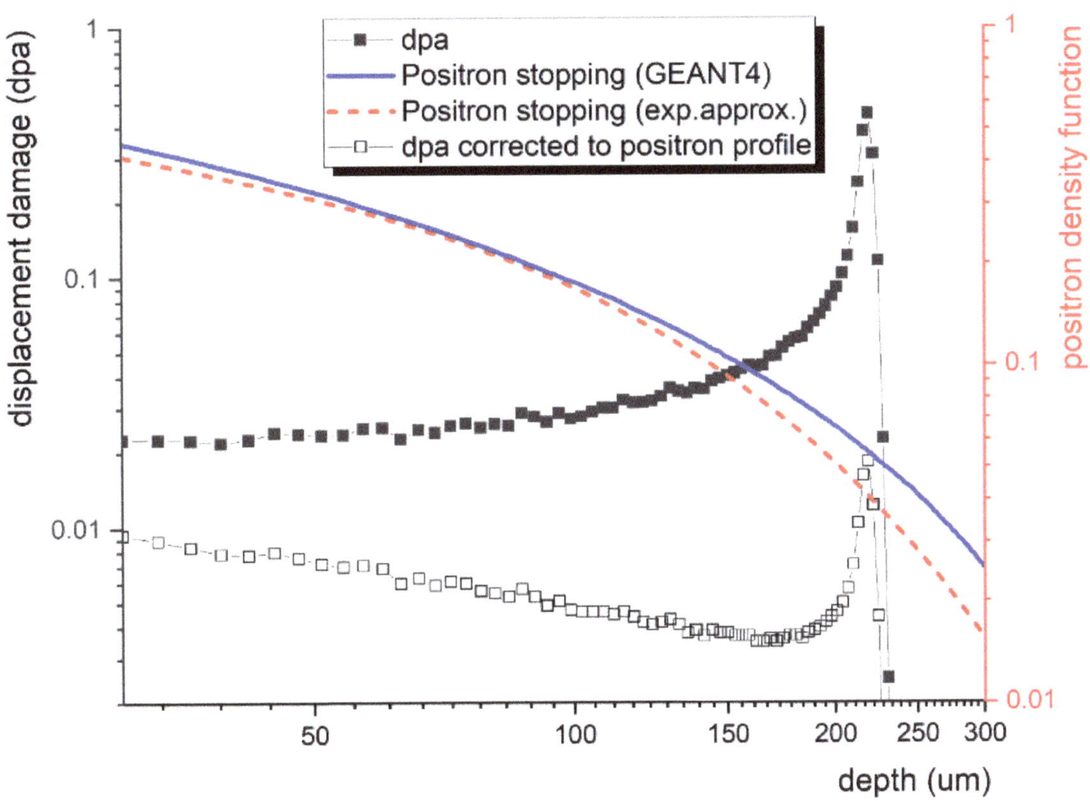

Figure 3. SRIM-based calculation for the dpa with 5 MeV protons.

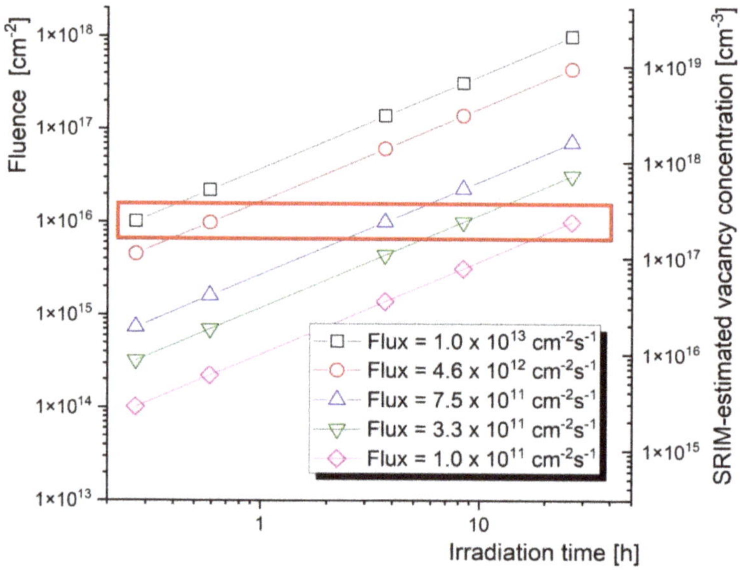

Figure 4. Fluence vs. irradiation times for different proton fluxes (beam currents) used in the experiment.

The irradiation temperature was kept near room temperature using a water-cooled sample stage. Figure 3 shows the simulated implantation profile and the values of displacement per atom (dpa) calculated according to the Norgett–Robinson–Torrens (NRT) [18] model using the "The Stopping and Range of Ions in Matter" (SRIM) data obtained according to the suggestion by Stoller [19] for the fluence of 10^{16} cm^{-2} used in the present study.

The fundamental approach in this study assumed that the concentration of the radiation-induced point defects would be constant at a certain range of proton flux, but there is a sharp threshold at a certain level of proton flux above which the production of these defects will be diminished in the thermal effects produced by the dislocation cascades.

The SRIM code was used to compute the reference value of the produced concentration of radiation-induced vacancies. However, it is a common understanding that this tool does not consider the mobility of the displaced atoms and the results are obtained for 0 K temperature. Practically, there is always a temperature effect that reduces the number of actual concentrations of vacancies that survived the displacement cascades. Positron annihilation spectroscopy can effectively study this realistic assessment of the concentration of radiation-induced vacancies.

The energy of the charged particles plays a leading role in the type of defects, such as Frenkel pairs, as well as cascade and sub-cascaded collisions. In the presented experiments, energy of 5 MeV was used for the proton irradiation and the resulting sample modification. The corresponding SRIM profile is shown in Figure 3, together with the positron stopping profile obtained from the GEANT4 (GEometry ANd Tracking) simulation package [20]. The figure illustrates the sensitivity of this technique to the given (uneven) defect depth profile by providing the actual/corrected dpa profile, "visible" to ^{22}Na positrons. Energy of 5 MeV was chosen so as to minimise the interaction of positrons with the displacement damage peak and the hydrogen peak produced at the end of the track region by protons capturing electrons.

As mentioned above, the reasoning for selecting the 10^{16} cm^{-2} fluence is derived from Figure 4. This fluence can be obtained using realistic flux values and accelerator beam time availability.

2.3. Positron Annihilation Spectroscopy Characterisation

Among the numerous analytical techniques used in material irradiation studies, positron annihilation spectroscopy (PAS) is well known for its spectacular sensitivity to atomic-scale vacancy-type defects. Although the technique is sensitive to other types of defects (dislocations, grain boundaries and precipitates of certain elements), vacancy-type defects are typically the most attractive potential well in irradiated single crystals. The sensitivity of various PAS techniques to neutral vacancies ranges from ~5×10^{15} cm^{-3} (detection limit) to ~10^{19} cm^{-3} (positron trapping gets saturated). This sensitivity range was also considered in the selection of the target fluence in the present experiment.

The experimental characterisation of the irradiated samples was performed at the Slovak University of Technology in Bratislava at the Faculty of Electrical Engineering and Information Technology at the Institute of Nuclear and Physical Engineering. This institute has a dedicated PAS laboratory for positron annihilation spectroscopy equipped with one standalone positron lifetime spectrometer and one setup combining positron lifetime and coincidence Doppler broadening spectrometer. Both lifetime spectrometers are digital, based on three B_aF_2 scintillator detectors and DRS4 waveform digitising boards.

As mentioned above, the experiment was designed for optimal utilisation of a conventional ^{22}Na positron source with a continuous energy spectrum of positrons ranging from 0 to 540 keV. The actual positron stopping profile, as well as the displacement damage profile adjusted to the spectrum of positron probes, is shown in Figure 3.

For the present research, we used positron annihilation lifetime spectroscopy (PALS), which enables qualitative and quantitative characterisation of vacancy-type defects in crystalline materials. The physical principle of the PALS technique is based on positron trapping by defects and the fact that the positron lifetime depends on the nature and size of this defect. PALS is a widely used microstructural characterisation technique based on the measurement of changes in the time of positrons trapped by lattice defects. Both the size and concentration of defects can be obtained from the positron lifetime spectrumby evaluating the lifetime values and intensities of individual components. The values of the positron lifetimes are well-known for most semiconductors, and the evaluation of the results can be supported by a broad range of published data that are both theoretically and experimentally obtained.

3. Results and Discussion

The positron annihilation lifetime spectra were evaluated using the LT10 program developed by Giebel and Kansy [21]. The spectra were decomposed into two components. The first component characterises the material bulk lifetime t_1 (reduced by trapping at defects) and the second component corresponds to lattice defects t_2, here considered as monovacancies. The lifetime of the second component was fixed at a value of 295 ps, reported for a mono-vacancies in undoped GaAs [22], while the lifetime of the first component, together with both intensities (I_1, I_2), was left as a free parameter. The average positron lifetime (t_{AVG}), as the statistically most reliable parameter independent of the fitting model, was calculated for all of the lifetime data. The concentration of vacancies was calculated according to the procedure described in detail, for instance, in [23]. The obtained results are shown in Table 3. The concentration of vacancies N_V was directly calculated from the positron trapping rate k_v via a constant of proportionality, the so-called trapping coefficient of 1×10^{15} s^{-1} [22].

The results plotted in Figure 5 show that the proton flux ranges between 10^{11} and 10^{12} $cm^{-2}.s^{-1}$ lead to about the same concentration of monovacancies in $5 \pm 1 \times 10^{16}$ cm^{-3}. This value is fairly reasonable compared with the SRIM simulation, which estimates the vacancy concentration for the given fluence on the level 2.56×10^{17} cm^{-3}. The discrepancy is given by the fact that SRIM does not account for the thermal recombination of vacancies, so SRIM always overestimates the actual damage to the lattice. Considering the aim of this paper, it is interesting to compare the observed vacancy production with other types of irradiation experiments involving PALS analysis. The paper by Sagatova et al. [24] on 8

MeV electron irradiated GaAs reported vacancy concentrations of 1.6 and 2.8 × 10^{16} cm^{-3} for samples exposed to 1000 and 1500 kGy radiation. As the latter dose was obtained by 8.37 × 10^{15} cm^{-2} electron fluence, i.e., close to the proton fluence 1 × 10^{16} cm^{-2} reported in this paper, one can compare the impact of two different types of radiation. Such a comparison suggests that these two types of radiation introduce a similar resulting displacement damage, with only a slightly higher (~factor of 2) concentration of vacancies produced by protons. It is important to note that the analysis was aimed at the ion track region and not the damage peak in both cases.

Table 3. Experimental results of the PAS.

GaAs Samples	p^+ Flux [cm^{-2}]	t_1 [ps]	I_1 [%]	t_2 [ps]	I_2 [%]	t_{AVG} [ps]	FV	k_V [s^{-1}]	N_V [cm^{-3}]
Set 1	1.04 × 10^{13}	223	98.98%	295	1.02%	223.7	0.97	9.27 × 10^6	4.64 × 10^{14}
Set 2	4.63 × 10^{12}	221	85.00%	295	15.00%	231.8	0.93	1.58 × 10^8	7.92 × 10^{15}
Set 3	7.51 × 10^{11}	215	50.00%	295	50.00%	254.0	1.05	8.97 × 10^8	4.49 × 10^{16}
Set 4	3.30 × 10^{11}	210	50.00%	295	50.00%	251.5	0.99	8.97 × 10^8	4.49 × 10^{16}
Set 5	1.04 × 10^{11}	203	46.00%	295	54.00%	251.6	1.07	1.05 × 10^9	5.27 × 10^{16}

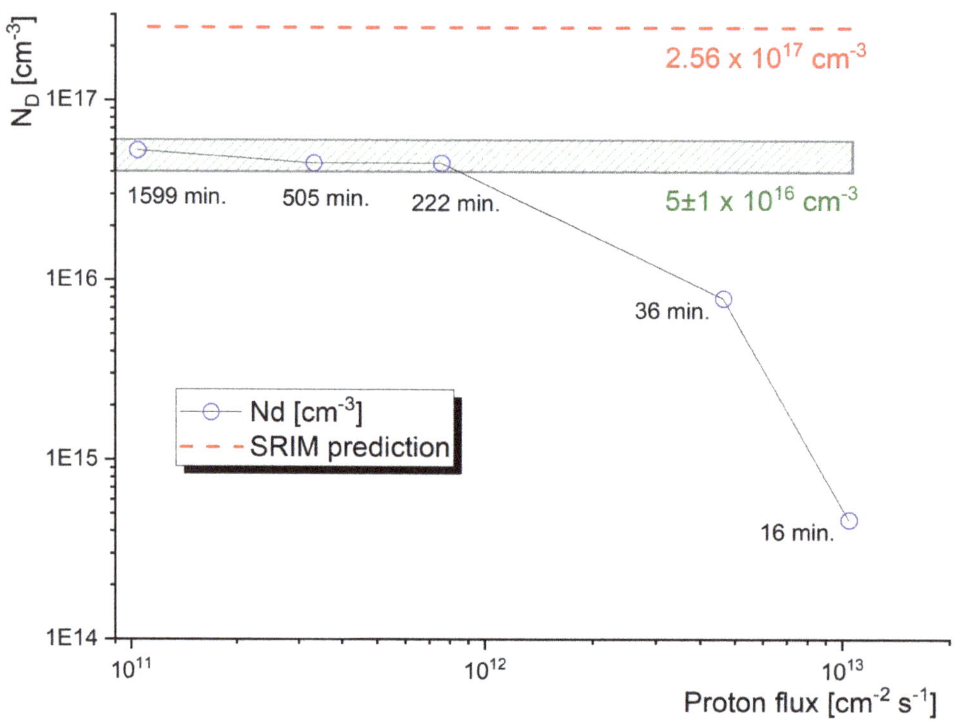

Figure 5. The concentration of radiation-induced vacancies in the studied GaAs samples as obtained from the PALS experiments. The concentration predicted by SRIM for 0 K is also indicated.

As can be further seen from Figure 5, at a certain level of proton flux (> 10^{12} s^{-1} cm^{-3}) the concentration of radiation-induced vacancies dropped sharply, suggesting that the new displacement damage cascades were initiated while the previous cascades were still active.

On the other hand, it is reasonable to assume that proton fluxes lower than ~10^{11} s^{-1} m^{-3} would lead to a defect concentration near the saturation range indicated in the figure. From this, we can conclude that proton flux below ~10^{12} s^{-1} cm^{-3} provides a meaningful radiation condition for the accelerated ageing studies in semiconductors planned for application in harsh radiation environments.

This is an important observation for numerous future experiments as it suggests that long exposure to a mild radiation environment can be, to some extent, simulated experimentally by short-term exposure to much more severe radiation conditions. Such a significant shortening of the irradiation experiment can significantly save costs related to beamtime at irradiation facilities.

It is important to note that the observed flux effect could be very different from the flux effect reported for neutron irradiation experiments on reactor pressure vessel (RPV) steels and other types of complex materials, where a higher flux leads to a more significant vacancy-type defect production [25]. Unlike the present experiment, the microstructure of irradiated RPV steels suffers from additional segregation and precipitation of certain elements (such as Cu or P), which the radiation-induced vacancies may be associated with. While the microstructural evolution of semiconductors is relatively simple compared with the nuclear structural materials, the number of published reports on the flux effect in these materials is significantly smaller.

In this study, the electrical properties of the semiconductors were not investigated, but it is reasonable to assume that the concentration of the free charge carriers would result in similar conclusions. However, this will be investigated in more detail in our forthcoming study, additionally including wide-bandgap semiconductors.

4. Conclusions

The present study reports an experimental quantitative characterisation of radiation-induced vacancies in GaAs obtained in proton irradiation experiments using a wide range of proton fluxes. The experimental data were obtained by positron annihilation lifetime spectroscopy, considering the actual stopping profiles of both projectile (proton) and probe (positron) particles. The present research can be summarised as follows:

- Positron annihilation spectroscopy can be effectively used as a tool for the quantitative characterisation of vacancy-type defects in semiconductors exposed to harsh radiation environments. While the present experiments led to a vacancy concentration near the saturation limit of PAS, the optimal proton fluence for future irradiation experiments can be selected from the range of 10^{15}–10^{16} cm^{-2}.
- The results indicate that mild radiation environments involving high-energy protons can be effectively simulated and accelerated by employing relatively high proton fluxes. Moreover, the proton irradiation seems to induce a concentration of vacancy-type defects (mono-vacancies) that is reasonably similar to high-energy electron irradiation experiments with a similar fluence.
- While SRIM code simulations provide data about the production rates of radiation-induced defects, the presented PAS characterisation enables reliable quantification of the survival rate of the defects. Similar to numerous studies in the past and referenced in this work, the presented experiment can be expanded to include the study of the recovery of the microstructure after thermal annealing of the samples.
- There is a threshold flux above which the proton irradiation experiment becomes unreasonable and inefficient. This threshold is relatively high and lies above 10^{12} s^{-1} cm^{-2}. At a higher proton flux, the new displacement damage cascades are initiated while the previous cascades are still occurring. This results in a sharp reduction in the concertation of the surviving vacancies.

In the next experiment, these conclusions will be used for proposing irradiation studies on other types of semiconductors, including wide bandgap semiconductors. A planned combination of PAS experiments and measurements of the electrical properties of the irradiated materials, such as resistivity and free charge carriers' concentration, will potentially increase the knowledge about the radiation tolerance of WBG materials, which are inevitable in numerous applications, including space exploration and safety for nuclear power installations.

Author Contributions: Conceptualization, I.N. and V.K.; methodology, I.N., V.K. and P.N.; irradiation experiment P.N.; software, M.P.; investigation, M.P. and S.S.; data curation, I.N. and V.K.; writing—original draft preparation, I.N. and V.K.; writing—review and editing, V.S., P.N. and J.D.; supervision, V.K.; project administration, V.K, J.D., P.N. and V.S.; funding acquisition, P.N. and V.S. All authors have read and agreed to the published version of the manuscript.

Funding: The authors would like to acknowledge partial support from the Slovak Research and Development Agency grants No. APVV-20-0010, APVV-20-0220 as well as from the European Regional Development Fund, projects No. ITMS2014+: 313011BUH7 and ITMS2014+: 313011W085. The authors further acknowledge financial contributions from the Scientific Grant Agency of the Ministry of Education, Science, Research and Sport of the Slovak Republic and the Slovak Academy of Sciences, grant number VEGA 1/0395/20.

Institutional Review Board Statement: Not applicable.

Informed Consent Statement: Not applicable.

Data Availability Statement: All data used to reach the conclusions are presented in the paper. Raw data and experimental logs are available upon request.

Conflicts of Interest: The authors declare no conflict of interest.

References

1. Van Tuyl, R.L. The Early Days of GaAs Ics. In Proceedings of the 2010 IEEE Compound Semiconductor Integrated Circuit Symposium (CSICS), Monterey, CA, USA, 3–6 October 2010; pp. 1–4. [CrossRef]
2. Pearton, S.J.; Aitkaliyeva, A.; Xian, M.; Ren, F.; Khachatrian, A.; Ildefonso, A.; Kim, J. Review—Radiation Damage in Wide and Ultra-Wide Bandgap Semiconductors. *ECS J. Solid State Sci. Technol.* **2021**, *10*, 055008. [CrossRef]
3. Reed, K.; Goetz, C.; Ericson, N.; Sweeney, D.; Ezell, N.D. *Wide Bandgap Semiconductors for Extreme Temperature and Radiation Environments;* Oak Ridge National Lab: Oak Ridge, TN, USA, 2022. [CrossRef]
4. Kizilyalli, I.C.; Xu, Y.A.; Carlson, E.; Manser, J.; Cunningham, D.W. Current and Future Directions in Power Electronic Devices and Circuits Based on Wide Band-Gap Semiconductors. In Proceedings of the 2017 IEEE 5th Workshop on Wide Bandgap Power Devices and Applications (WiPDA), Albuquerque, NM, USA, 30 October–1 November 2017; p. 417. [CrossRef]
5. Huang, A.Q. Wide bandgap (WBG) Power Devices and their Impacts on Power Delivery Systems. In Proceedings of the 2016 IEEE International Electron Devices Meeting (IEDM), San Francisco, CA, USA, 3–7 December 2016; pp. 20.1.1–20.1.4. [CrossRef]
6. Shenai, K. Future Prospects of Widebandgap (WBG) Semiconductor Power Switching Devices. *IEEE Trans. Electron. Devices* **2015**, *62*, 248–257. [CrossRef]
7. Wide-Bandgap Semiconductors for Space Applications; Last Update: 15 February 2022. Record Number: 96633; European Union. 2022. Available online: https://cordis.europa.eu/project/id/242394 (accessed on 1 December 2022).
8. Millán, J.; Godignon, P.; Perpiñà, X.; Pérez-Tomás, A.; Rebollo, J.A. Survey of Wide Bandgap Power Semiconductor Devices. *IEEE Trans. Power Electron.* **2014**, *29*, 2155–2163. [CrossRef]
9. Maurer, R.H.; Fraeman, M.E.; Martin, M.N.; Roth, D.R. Harsh Environments: Space Radiation Environment, Effects, and Mitigation. *John Hopkins APL Tech. Dig.* **2008**, *28*, 17.
10. Bourban, G. Radiation Environment and its Effects in EEE Components and Hardness Assurance for Space Applications. In Proceedings of the CERN-ESA-SSC workshop European Space Agency ESA, Zurich, Switzerland, 9–10 May 2017.
11. Krstic, M. A Methodology for Characterization, Modeling and Mitigation of Single Event Transient Effects in CMOS Standard Combinational Cells. Doctoral Thesis, Postdam University, Postdam, Germany, April 2022. [CrossRef]
12. Rafí, J.M.; Pellegrini, G.; Godignon, P.; Ugobono, S.O.; Rius, G.; Tsunoda, I.; Moll, M. Electron, Neutron, and Proton Irradiation Effects on SiC Radiation Detectors. *IEEE Trans. Nucl. Sci.* **2020**, *67*, 2481–2489. [CrossRef]
13. Summers, G.P.; Burke, E.A.; Shapiro, S.R.; Walters, R.J. Damage correlations in semiconductors exposed to gamma, electron and proton radiations. *IEEE Trans. Nucl. Sci.* **1993**, *40*, 1372–1379. [CrossRef]
14. Marshall, P.W.; Dale, C.J.; Summers, G.P.; Wolicki, E.A.; Burke, E.A. Proton, neutron, and electron-induced displacement damage in germanium. *IEEE Trans. Nucl. Sci.* **1989**, *36*, 1882–1888. Available online: https://www.osti.gov/biblio/7202239 (accessed on 13 December 2022). [CrossRef]
15. Väyrynen, S. *Irradiation of Silicon Particle Detectors with Mev-Protons;* University of Helsinki Report Series in Physics: Helsinky, Finland, 2010; ISBN 978-952-10-5975-9.
16. Advanced Technologies Research Institute. *B8405—Specifications and Machine Logs for 6 MV Tandetron System;* Internal document of the ATRI MTF STU; Advanced Technologies Research Institute: Tokyo, Japan, 2022.

17. Noga, P.; Száraz, Z.; Kubiš, M.; Dobrovodský, J.; Ferenčík, F.; Riedlmajer, R.; Krsjak, V. High-Fluence Multi-Energy Ion Irradiation for Testing of Materials. *Materials* **2022**, *15*, 6443. [CrossRef] [PubMed]
18. Norgett, M.J.; Robinson, M.T.; Torrens, I.M. A proposed method of calculating displacement dose rates. *Nucl. Eng. Des.* **1975**, *33*, 50. [CrossRef]
19. Stoller, R.E. On the use of SRIM for computing radiation damage exposure. *Nucl. Instrum. Methods Phys. Res. Sect. B: Beam Interact. Mater. At. Vol.* **2013**, *310*, 75–80. [CrossRef]
20. Allison, J.; Amako, K.; Apostolakis, J. Geant4 developments and applications. *IEEE Trans. Nucl. Sci.* **2006**, *53*, 270–278. [CrossRef]
21. Giebel, G.; Kansy, J. LT10 Program for Solving Basic Problems Connected with Defect Detection. *Phys. Procedia.* **2012**, *35*, 122–127. [CrossRef]
22. Rehberg, R.K.; Leipner, H.S. *Positron Annihilation in Semiconductors*; Springer Verlag: Berlin/Heidelberg, Germany, 1999; p. 378. [CrossRef]
23. Krsjak, V.; Degmova, J.; Noga, P.; Petriska, M.; Sojak, S.; Saro, M.; Slugen, V. Application of Positron Annihilation Spectroscopy in Accelerator-Based Irradiation Experiments. *Materials* **2021**, *14*, 6238. [CrossRef] [PubMed]
24. Sagatova, A.; Krsjak, V.; Sojak, S.; Riabukhin, O.; Kovacova, E.; Zatko, B. Semi-insulating GaAs detectors degraded by 8 MeV electrons up to 1500 kGy. *J. Instrum.* **2021**, *16*, C12032. [CrossRef]
25. Ulbricht, A.; Hernández-Mayoral, M.; Oñorbe, E.; Etienne, A.; Radiguet, B.; Hirschmann, E.; Bergner, F. Effect of Neutron Flux on an Irradiation-Induced Microstructure and Hardening of Reactor Pressure Vessel Steels. *Metals* **2022**, *12*, 369. [CrossRef]

materials MDPI

Article

Evaluation of Radiation Resistance of Polystyrene Using Molecular Dynamics Simulation

Yeong-Heum Yeon †, Ha-Eun Shim †, Jin-Hyung Park, Nam-Ho Lee, Jae-Yeon Park, Moon-Sik Chae, Jung-Ho Mun, Jae-Hyun Lee and Hui-Jeong Gwon *

Advanced Radiation Technology Institute, Korea Atomic Energy Research Institute, 29 Geumgu-Gil, Jeongeup-Si 56212, Korea; yhyeon@kaeri.re.kr (Y.-H.Y.); she0805@kaeri.re.kr (H.-E.S.); jhpak@kaeri.re.kr (J.-H.P.); nhlee@kaeri.re.kr (N.-H.L.); jaeyeon@kaeri.re.kr (J.-Y.P.); cmswill@kaeri.re.kr (M.-S.C.); jhmun@kaeri.re.kr (J.-H.M.); jaehyunlee@kaeri.re.kr (J.-H.L.)

* Correspondence: hjgwon@kaeri.re.kr; Tel.: +82-063-570-3087

† These authors equally contributed to this work.

Abstract: The scission rates of polystyrene and fluorinated polystyrene irradiated in an irradiation facility with Co-60 γ-rays were determined using molecular dynamics simulation and gel permeation chromatography (GPC) molecular weight distributions. The prediction was based on the assumption that γ-ray energy is transferred to the initial velocity of the primary knock-on atom. We employed a molecular dynamics simulation procedure to compute the changes in bond length between the connections for selected values of the absorbed dose and compared the calculated values with measurements made on the irradiated samples. The samples were exposed to four different absorbed doses of 25, 50, 75, and 100 kGy. The scission process and scission ratio were simulated with LAMMPS with ReaxFF potential for each bond, and we compared the simulation results with the experimental data especially measuring average molecular weight to evaluate the effect of fluorination on radiation enhancement.

Keywords: gel permeation chromatography; scission; polystyrene; fluorinated polystyrene; molecular dynamics simulation; radiation resistance

Citation: Yeon, Y.-H.; Shim, H.-E.; Park, J.-H.; Lee, N.-H.; Park, J.-Y.; Chae, M.-S.; Mun, J.-H.; Lee, J.-H.; Gwon, H.-J. Evaluation of Radiation Resistance of Polystyrene Using Molecular Dynamics Simulation. *Materials* **2022**, *15*, 346. https://doi.org/10.3390/ma15010346

Academic Editor: Vladimir Krsjak

Received: 26 November 2021
Accepted: 21 December 2021
Published: 4 January 2022

1. Introduction

Polystyrene is a standard plastic used in a radiative environment for applications in the nuclear industry and nuclear reactors and outer space localization. Irradiation experiments have been used to confirm whether these plastics can be used as phantom materials to measure absorbed doses. These materials have been increasingly used in various radiation dosimetry protocols. Solid plastics, such as polystyrene, were used for low-energy radiation dosimetry by the IAEA (2000) International Code of Practice TRS-398 [1]. The mechanical properties of these materials were measured by increasing the radiation dose to evaluate these materials. Crosslinking polymers, sterilizing, and preserving medical equipment or food are the application of radiation modification as the result of this method. These polymers have been modified to improve their radiation enhancement properties through experimental trial and error, but this method takes a long time [2,3]. Many experimental methods have been studied to increase the radiation enhancement of polystyrene by fluorination [4,5] They are comparing zero-strength-time and tensile-strength mechanically, and measuring G-value, detecting free radicals by electron spin resonance chemically [6,7].

The irradiation resistance assessment of polymers has been comprehensively performed mechanically, including tensile strength and elongation at break, for tensile tests. Several studies have shown that the "equal dose–equal damage" concept is not appropriate because of the complexity of radiation reactions in polymers [8,9]. Polymerizing, grafting, chain scission, and crosslinking are the main reactions to irradiation. The study and simulation of irradiation effects focused on crosslinking because of its importance, particularly the improvement of mechanical and thermal properties of the polymer [10,11]. The types of irradiation that cause breaking chains are electrons, neutrons, α- and β-particles, or γ- and X-rays. The chemical and physical aging of polystyrene due to the γ–ray was studied, especially the effect of air on the radiation-induced changes in mechanical and molecular properties by considering scission and crosslinking yields, G(S) and G(X) [12,13]. The crosslinking reaction in polystyrene with γ-ray was studied especially deuterated styrene was studied [14]. However, crosslinking occurs by breaking the chains to produce radicals. These radicals are the main factors involved in the crosslinking of polymers. Therefore, we simulated microscopic information obtained by theoretical methods and molecular dynamics to predict the scission ratio of each bond in polystyrene and fluorinated polystyrene and compared the simulation data with experimental data especially measuring average molecular weight to evaluate the effect of fluorination on radiation enhancement.

In this study, we simulated the scission rates of polystyrene and fluorinated polystyrene and irradiated them in an irradiation facility with Co-60 γ-rays and compared the simulation results and average molecular weight measurement. Molecular dynamics methods calculated the polystyrene and fluorinated polystyrene bond lengths by increasing the absorbed dose to predict the scission rate and provide a radiation enhancement method for each polymer. The radiation resistance was investigated by the partial substitution of H atoms for F atoms.

2. Materials and Methods

2.1. Materials

The styrene monomer, 2,3,4,5,6-pentafluorostyrene, tetrahydrofuran and di-tert-butyl peroxide (DTBP) as an initiator for polymerization were purchased from Sigma-Aldrich Korea (Seoul, Republic of Korea). 4-fluorostyrene was purchased from Tokyo Chemical Industry Co., Ltd. (Tokyo, Japan). All commercially available reagents were used as received without further purification.

2.2. Determination of Average Molecular Weight Sample Preparation and γ Radiation Experiments

Polystyrene (PS) and fluorinated styrene (1F-, 5F-PS) were synthesized as per a procedure described in the literature with slight modifications (Figure 1) [15]. The styrene monomer (10.5 g, 0.1 mol) was suspended in 10 mL of tetrahydrofuran (THF) and 1.42 g (0.01 mol) of di-tert-butyl peroxide (DTBP) was added. The polymerization was carried out at 100 °C for 18 h, at which point the solution is cooled to room temperature (20 °C). The reaction product (PS) was precipitated in a solution of methanol (200 mL), vacuum-filtered, and dried at 60 °C for five days to evaporate the remaining solvent. To synthesize the fluorinated styrene, the same procedure as described above was used with 12.2 g (0.1 mol) of unit 4-fluorostyrene to the 1F-PS and 9.7 g (0.05 mol) of unit 2,3,4,5,6-pentfluorostyrene to the 5F-PS, respectively.

Prior to gamma irradiation, the prepared samples were packed in glass vials. After that, gamma irradiation was performed with respective irradiation doses (25, 50, 75, and 100 kGy) at a dose rate of 10 kGy/h. The Co-60 source (MDS Nordion, Ottawa, Ontario Canada, IR 221 n wet storage type C-188) was located at the Korea Atomic Energy Research Institute (KAERI), Jeongeup-si, Korea.

2.3. Determination of Average Molecular Weight

Changes in the molecular weights of the irradiated PS, 1F-PS, and 5F-PS were determined by gel permeation chromatography (GPC, Breeze system, Waters, Milford, USA) with the eluent of $CHCl_3$ at a flow rate of 1.0 mL/min at 40 °C. This equipment consisted of a Water 1515 Isocrylic HPLC pump, Water 2414 reflective index detector, Phenogel column 5 μm (300 × (4.6 mm)), and calibrated with polystyrene standards (Shodex, SM-105, and SL-105).

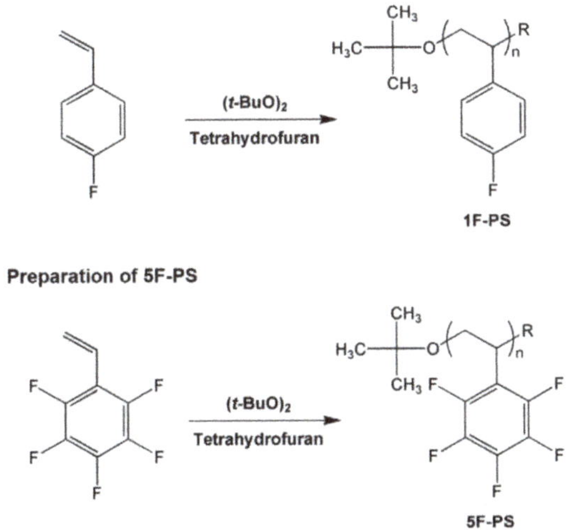

Figure 1. Preparation of Ps, 1F-PS(poly(4-fluorostyrene)) and 5F-PS(poly(2,3,4,5,6-pentafluorostyrene)).

3. Theoretical Study of Scissionning Reactions

3.1. Modeling

The structure of polystyrene and fluorinated polystyrene were drawn using Avogadro and optimized using LAMMPS [16,17]. We applied the reactive hydrocarbon potential AIREBO and optimized its structure at room temperature [18]. The simulation size was 40 × 40 × 40 cubic Å for one cell. The polystyrene in on cell was shown in Figure 2.

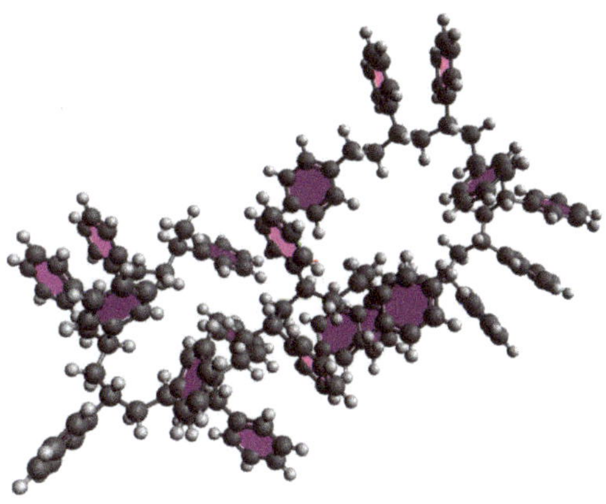

Figure 2. Polystyrene with 322 atoms.

We applied a reactive force field potential to consider the fluorine atoms [19,20]. The fluorinated polystyrene, 1F-PS(poly(4-fluorostyrene) and 5F-PS(poly(2,3,4,5,6-pentafluorostyrene)), are shown in Figure 3. The length of each atom was calculated and compared with the cut-off distance of each bond using the same method as that for polystyrene. We performed the simulation ten times in 10 cells under the same conditions.

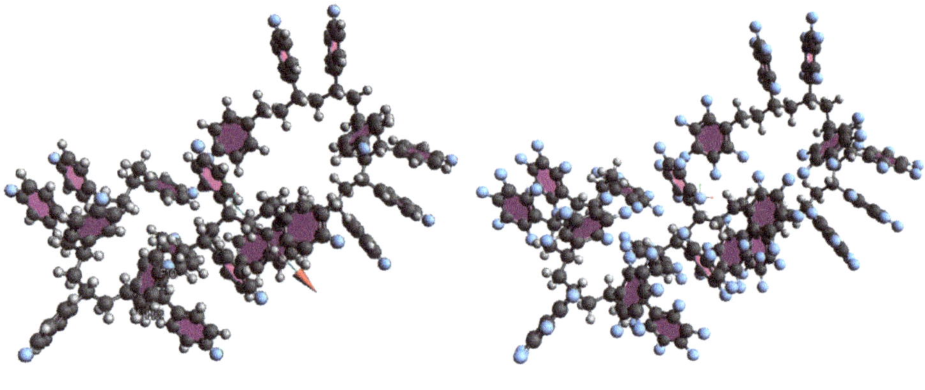

Figure 3. Fluorinated polystyrene: (**a**) 1F-PS(poly(4-fluorostyrene)) (**b**) 5F-PS(poly(2,3,4,5,6-pentafluorostyrene)).

3.1.1. Polystyrene

For polystyrene simulation, 322 atoms in 1 cell (total 10 cells) were drawn using Avogadro and are shown in Figure 2. We optimized the atoms in Avogadro, and converted the modeling file to the LAMMPS data file.

3.1.2. Fluorinated Polystyrene

For the fluorinated polystyrene simulation, we substituted the benzene ring of polystyrene with fluorine to evaluate the radiation enhancement. We optimized the atoms in Avogadro with the UFF force field and converted the modeling to the LAMMPS data file.

3.2. Reactive Molecular Dynamics Simulation

The scission rate was calculated using molecular dynamics simulations to evaluate the radiation resistance of polystyrene and fluorinated polystyrene. These materials were simulated using the LAMMPS code. We applied the reactive hydrocarbon potential AIREBO developed by Brenner for polystyrene and reactive force field potential (ReaxFF) developed by Van Duin et al. for fluorinated polystyrene [18,19].

High-energy particle irradiation has often been simulated using molecular dynamics codes such as LAMMPS and PARCAS, for instance, in the study of Beardmore et al. [20,21]. The high-energy particle irradiation to material is simulated by applying the interaction between the particles, the colliding electron, and the lattice atom. This method was studied by giving initial kinetic energy, the recoil energy, to some randomly chosen atom in the lattice [22–24].

However, this study simulated the interaction between γ-rays and atoms by providing γ-ray energy in a specific area of the chain of polymers. We converted the γ-ray energy to the velocity of the primary knock-on atom (PKA) that existed in a randomly chosen area (included recoil atom) and calculated the scission rate of each bond. This approach is suitable for investigating the scission rate of each bond during initial energy conversion.

The energy of the γ-ray was 1332 keV, and we converted it to a momentum term to calculate the velocity of the initial primary knock-on atom. The momentum of the initial primary knock on the atom was calculated using the following equation:

$$P = \frac{E}{c} \tag{1}$$

where c denotes the velocity of light. The initial velocity was calculated by the equation:

$$P = mv, \quad v = \frac{E}{mc} \tag{2}$$

where m is the mass of the initial primary knock-on atom. We assumed that all of the γ-ray energy converted to the kinetic energy of the primary knock-on atom. The calculated velocity was 274 Å/ps for the carbon atom. The direction of the momentum was the same direction as γ-ray.

The second assumption was that we selected the shape of the applied velocity area sphere where collision cascade occurs. The radius of the area was 1.6 Å, which is the C-C bond length. This area was increased by increasing the absorbed dose. We applied this assumption by the equation:

$$D_n = \frac{N_d}{N} = \int_0^t \int_0^\infty \int_{E_d}^{T_{max}} \varnothing(E,t) \frac{d\sigma(E,T)}{dt} \gamma(T) \mathrm{dTdEdt} \tag{3}$$

where D_n denotes the probability of displacement atoms, N denotes number density of the material, N_d denotes the number of displaced atoms per unit volume, t denotes irradiation time, $\varnothing(E,t)$ denotes flux of incident particle, $\sigma(E,t)$ denotes scattering cross-section, $\gamma(T)$ denotes the number of atoms displaced from origin position by PKA, E denotes the initial energy of the incident particles, T denotes transferred energy, E_d denotes the energy of the displaced atom and T_{max} is maximum energy transferred to PKA. In the simulation, the initial energy of the incident particle (E) was fixed by the value of conversion energy of γ-ray as the first assumption because the radiation intensity was set by the energy of γ-ray (1332 keV). The flux of incident particle was increased proportionally to the absorbed dose, so we increased the PKA region (collision cascade zone) proportional to the absorbed dose. Because increasing absorbed dose means that increasing the number of atoms displaced from origin position ($\gamma(T)$) is proportional to the flux of incident particles, including the irradiation time. The number of atoms displaced from the original position and their velocity by PKA are shown in Figure 4.

The absorbed dose of polystyrene and fluorinated polystyrene was 25, 50, 75, 100 kGy in the experiment. We increased the radius of the PKA area proportional to the absorbed dose, and they were 1.6, 2.3, 2.8, and 3.2 Å in the simulation. We optimized the geometry of each material, and the simulation time was one ps. We checked the distance of each atom at 10 cells after the 1 ps-the collision process and compared them with the bond break length (1.8, 1.09, and 1.35 Å for C-C, C-H, and C-F bond) to obtain the scission rate for each bond.

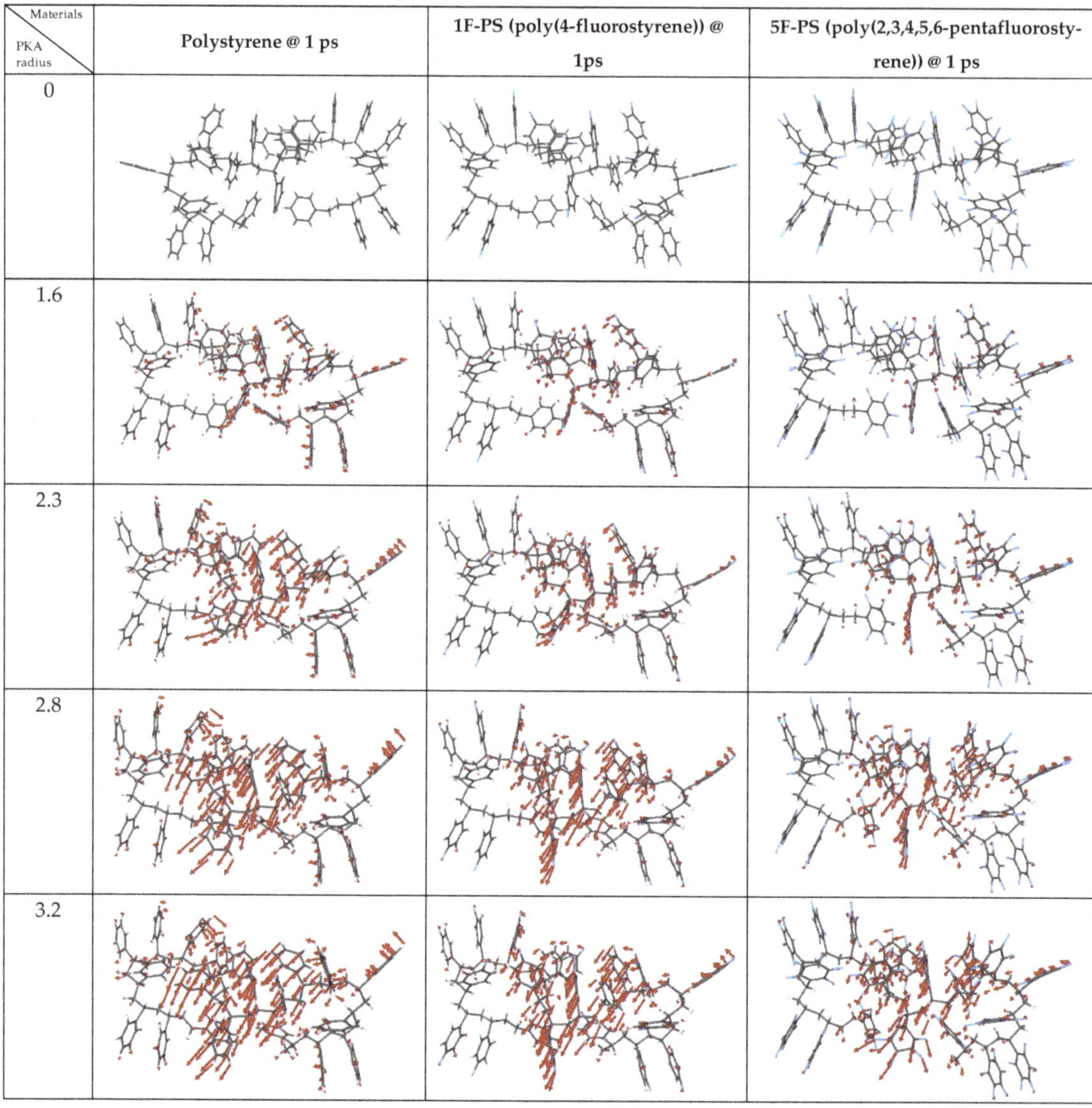

Figure 4. Snapshots of the displaced atoms' velocity vector (conversion of γ-ray energy to kinetic energy-collision in the cascade zone) at the polymers @ 1 ps with increasing the PKA radius.

4. Results

4.1. Simulation Results

The bond break lengths of each bond are 1.8, 1.09, 1.35 Å, corresponding to C-C, C-H, and C-F bonds, respectively. We obtained the distance of each bond at 1 ps after applying the velocity, which is shown in Figure 4. The maximum number of bond dissociations (@ 1 ps) for polystyrene and fluorinated polystyrene are listed in Tables 1–3.

Table 1. Maximum dissociation number of each bond in polystyrene @ 1 ps.

PKA Radius (Å)	Absorbed Dose (kGy)	No. of Simulations	Maximum Dissociation Number of C-C Bond	Maximum Dissociation Number of C-H Bond	Total Dissociation Number	Dissociation Rate (#/kGy)
1.6	25	10	7	35	42	1.68
2.3	50	10	50	71	117	3
2.8	75	10	67	78	140	0.92
3.2	100	10	70	87	157	0.68

Table 2. Maximum dissociation number of each bond in 1F-PS (poly(4-fluorostyrene)) @ 1 ps.

PKA Radius (Å)	Absorbed Dose (kGy)	No. of Simulations	Maximum Dissociation Number of C-C Bond	Maximum Dissociation Number of C-F Bond	Maximum Dissociation Number of C-H Bond	Total Dissociation Number	Dissociation Rate (#/kGy)
1.6	25	10	3	3	30	36	1.44
2.3	50	10	31	6	43	85	1.96
2.8	75	10	46	8	73	127	1.68
3.2	100	10	57	9	74	140	0.52

Table 3. Maximum dissociation number of each bond in 5F-PS (poly(2,3,4,5,6-pentafluorostyrene)) @ 1 ps.

PKA Radius (Å)	Absorbed Dose (kGy)	No. of Simulations	Maximum Dissociation Number of C-C Bond	Maximum Dissociation Number of C-F Bond	Maximum Dissociation Number of C-H Bond	Total Dissociation Number	Dissociation Rate (#/kGy)
1.6	25	10	0	0	0	0	0
2.3	50	10	16	15	15	47	1.88
2.8	75	10	45	38	35	118	2.84
3.2	100	10	53	43	37	133	0.6

The dissociation rate was calculated at the maximum dissociation number. The maximum dissociation rate region was 3 (#/kGy) from 25 to 50 kGy at polystyrene, 1.96 (#/kGy) from 25 to 50 kGy at the 1F-PS (poly(4-fluorostyrene)), and 2.84 (#/kGy) from 50 to 75 kGy at the 5F-PS (poly(2,3,4,5,6-pentafluorostyrene)).

The average dissociation number of each bond at 1 ps in polystyrene for 10 simulations is shown in Figure 5. The dissociation of C-C bonds is less than that of C-H bonds in polystyrene, implying that the scission rate of the backbone in polystyrene is less than that of other bonds.

In 1F-PS (poly(4-fluorostyrene)), the dissociation number of C-C bonds is also less than that of C-H and C-F bonds. The dissociation of C-F bonds occurred, as shown in Figure 6.

The average dissociation number of each bond at 1 ps in 5F-PS (poly(2,3,4,5,6-pentafluorostyrene)) is shown in Figure 7. Dissociation did not occur up to 25 kGy. The dissociation number of C-C bonds was higher than that of C-F and C-H bonds.

We did not compare the dissociation number of each bond in the polymers because the total number of C-H and C-F bonds are different for each polymer. However, the total dissociation number was comparable for radiation resistance according to the same atom number. The 5F-PS (poly(2,3,4,5,6-pentafluorostyrene)) had the lowest dissociation number, as shown in Figure 8.

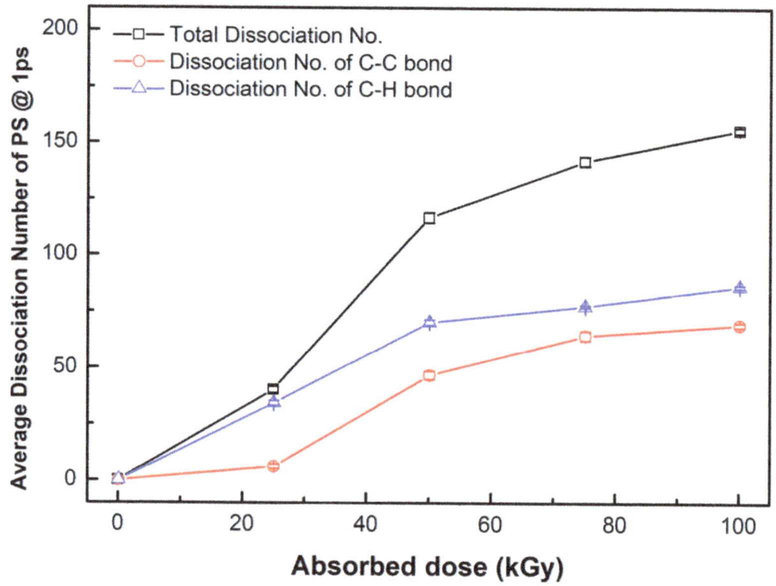

Figure 5. Average dissociation number of each bond in polystyrene at 1 ps.

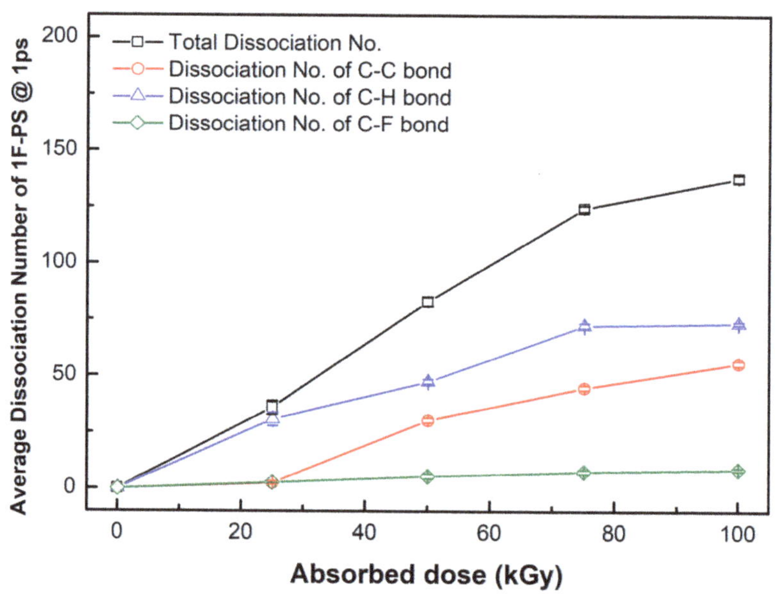

Figure 6. Average dissociation number of each bond in 1F-PS (poly(4-fluorostyrene)) at 1 ps.

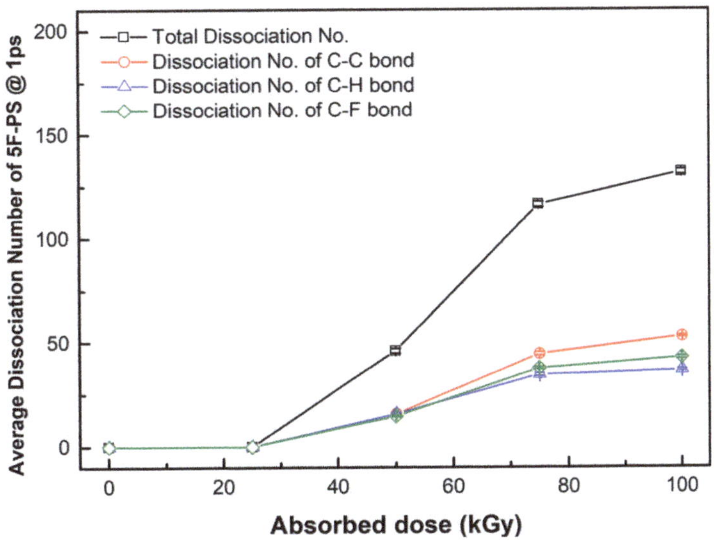

Figure 7. Average dissociation number of 5F-PS (poly(2,3,4,5,6-pentafluorostyrene)) at 1 ps.

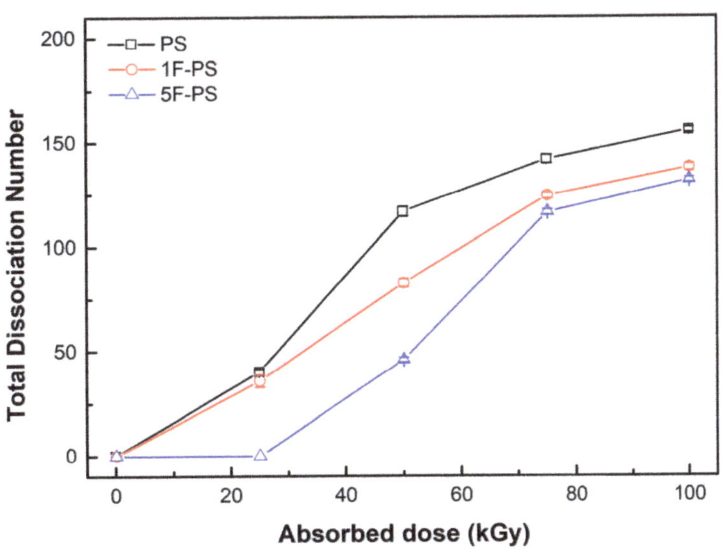

Figure 8. Total dissociation number of polymers.

4.2. Comparison with Experiment Results

We applied the scission rate by considering the mass of each atom for comparison with the gel permeation chromatography results. The results of polymers' average molecular weight (Mn) are shown in Figure 9, and the ratio of differential average molecular weight (Mn) of the irradiated polymers are summarized in Table 4. The values of Mns were different for each polymer. We applied the ratio of changes to compare the radiation resistance of each polymer. The ratio was calculated by standardizing the initial Mns value.

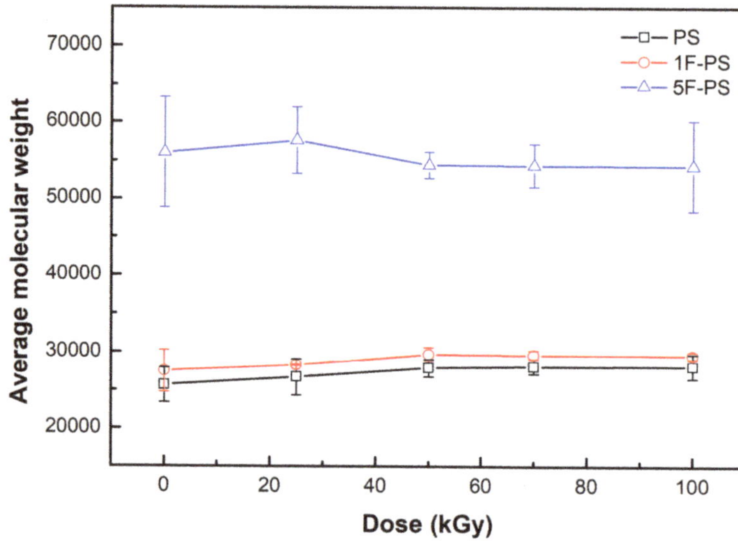

Figure 9. The polymers average molecular weight after irradiation (experiment results).

Table 4. Maximum dissociation number of each bond in polystyrene @ 1ps.

Absorbed Dose (kGy)	Mn of PS	Mn of 1F-PS	Mn of 5F-PS	The Ratio of Differential Mn (PS)	The Ratio of Differential Mn (1F-PS)	The Ratio of Differential Mn (5F-PS)
0	25,651.0	27,468.3	56,011.0	0.00000	0.00000	0.00000
25	26,714.3	28,249.0	57,630.0	0.02887	0.02032	0.01229
50	27,917.6	29,691.5	54,441.0	0.05890	0.05505	0.01261
75	28,089.0	29,639.0	54,391.0	0.06296	0.05384	0.01303
100	28,129.0	29,540.0	54,345.5	0.06391	0.05156	0.01340

In the process of the experiment (measuring average molecular weight by gel permeation chromatography after irradiation), both the scission and crosslinking processes had happened. The initial average molecular weights of polymers are 25,651.0, 27,468.3, 56,011.0 for PS, 1F-PS, and 5F-PS. The average molecular weight of PS increased to 28,129.0 at 100 kGy. In the case of polystyrene fluoride, the average molecular weight of 1F-PS increased to 29,691.5 at 50 kGy, decreased to 29,540.0 at 100 kGy, and the average molecular weight of 5F-PS increased to 57,630.0 at 25 kGy and decreased to 54,345.5 at 100 kGy. However, the scission process should be done before crosslinking process. So we assumed that the change of average molecular weight was dominant by the scission process. We normalized the difference of average molecular weight by matching the initial average molecular weight to zero, and then we calculated the absolute difference between the initial average molecular weight and after the irradiation as shown in Figure 10 [25–27]. We compared these differential values with the simulation results of the scission rate. Equations (4) and (5) are being applied for comparison.

$$\alpha\ (scission\ rate) = \frac{N}{m} \tag{4}$$

$$\delta(Ratio\ of\ Differential\ Mn) = \left(\frac{1}{M_0} - \frac{1}{Mn}\right) \times N \tag{5}$$

where m denotes the mass of bonds, Mn denotes the number average molecular weight after irradiation, M_0 denotes the initial average molecular weight, and N is the differential number: the ratio between the average molecular weight of atoms (experiment results) and atom number (simulation value) after irradiation. The scission rates of the bonds which were simulated in the polymers are shown in Figures 11–13.

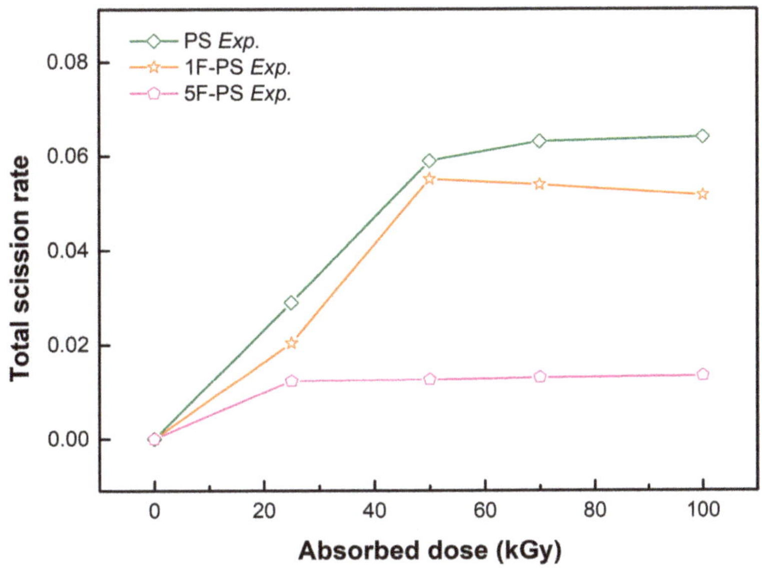

Figure 10. The polymers normalized average molecular weight after irradiation (experiment results).

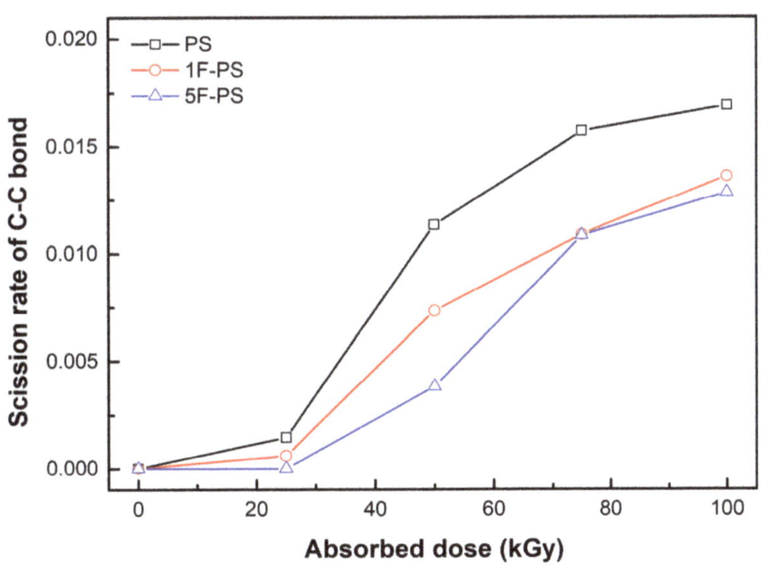

Figure 11. Scission rate of C-C bond of polymers.

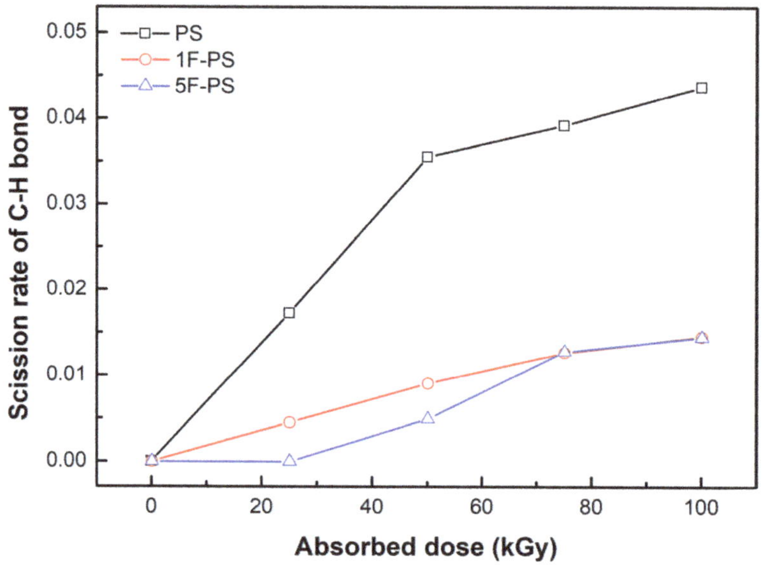

Figure 12. Scission rate of C-H bond of polymers.

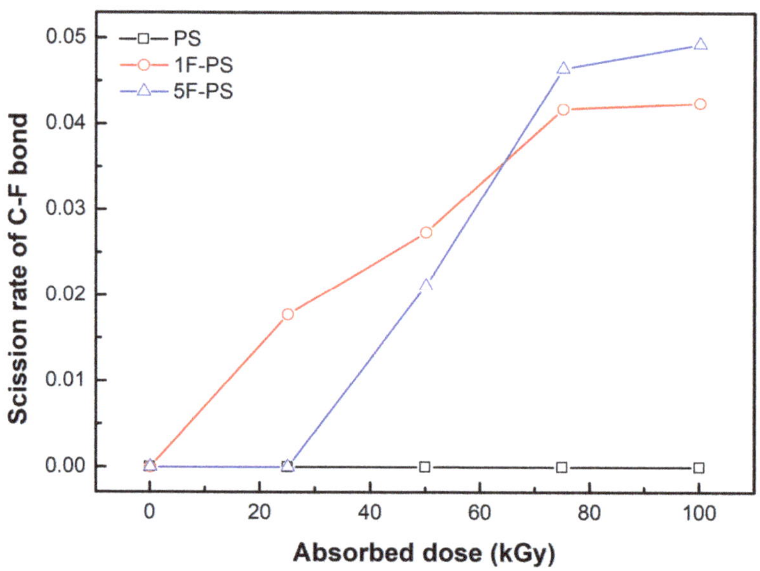

Figure 13. Scission rate of C-F bond of polymers.

The scission rate of the C-C bond was the highest in the polystyrene. The value of the scission rate was increased sharply in the region of 25 to 50 kGy for polystyrene and 1F-PS and 50 to 75 kGy for 5F-PS. These results were similar with the maximum dissociation rate region, which were 3 (#/kGy) from 25 to 50 kGy at the polystyrene, 1.96 (#/kGy) from 25 to 50 kGy at the 1F-PS, and 2.84 (#/kGy) from 50 to 75 kGy at the 5F-PS.

This relation was also investigated in the gel permeation chromatography results. The difference of Mn was the highest in the region of 25 to 50 kGy for polystyrene and 1F-PS.

However, it is different in the 5F-PS case. This means that the dominant scission happened in the C-C bond for polystyrene and 1F-PS.

The scission rate of the C-H bond was also the highest in the polystyrene. The value of the scission rate was increased linearly in the region of 0 to 50 kGy for polystyrene and 1F-PS and 75 to 100 kGy for 5F-PS. The increment of the scission rate was decreased after 50 kGy for polystyrene and 75 kGy for 1F-PS, 5F-PS.

The scission rate of the C-F bond in the 1F-PS was higher than 5F-PS until 75 kGy. The scission process did not happen until 25 kGy in the 5F-PS. The number of C-F bonds in 5F-PS is 5 times larger than 1F-PS. However, the scission rate of C-F bonds in 5F-PS was similar to 1F-PS.

The total scission rate of PS was higher than 1F-PS, 5F-PS in the simulation. The results of the scission trend of PS, 1F-PS was similar with the measured value, particularly until 50 kGy. The measured value was decreased after 50 kGy.

In the case of 5F-PS simulation, the scission did not happen until 25 kGy in the simulation. However, the measured value was increased until 25 kGy, and the rate of increase was the minimum in the data. The simulation results were compared with the experimental results and shown in Figure 14.

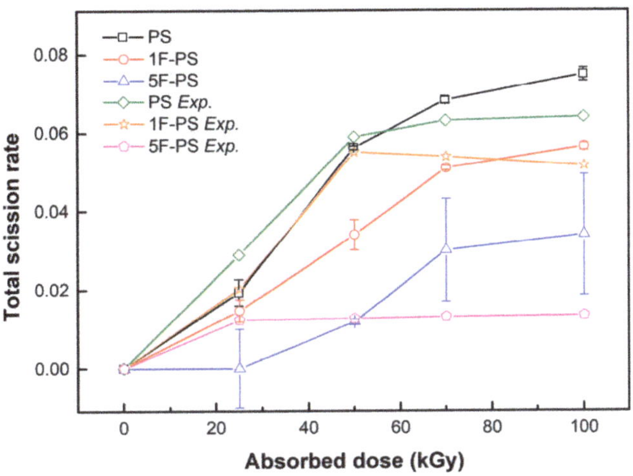

Figure 14. Total Scission Rate.

5. Discussion

The radiation enhancement effect with fluorine in the polystyrene has been determined by atomistic simulations to predict the possibility of using this method for modeling and simulation, which is irradiation reactions that occur in these materials.

The fluorinated PS backbone is made up of carbon–carbon bonds and the pendant groups are carbon-fluorine bonds. Both are extremely strong bonds. The basic properties of fluoropolymers from these two very strong chemical bonds. The quantity of the fluorine atom can be affected the crystallinity of PS and continuous covering around the carbon–carbon bonds and protect them from chemical attack, thus imparting chemical resistance and stability to the molecule.

In these polymers, chain scission (scission of C-C) was the primary reaction. The number of dissociation was increased proportionally with the absorbed dose. The ratio of increment of atoms in the simulation was the same with the measured (Mn) value.

5F-PS was found to be more resistant to radiation damage than 1F-PS and polystyrene. The benzene ring can be considered to act as a very efficient trap for F atoms due to the so-called cage effect. So, the F atoms are formed in an efficient cage of benzene molecules; the reactions (6) and (8) can predominate over reaction (9).

$$\sim C_6F_5 \rightarrow \sim C_6F_4\bullet + F\bullet \quad (6)$$

$$\sim C_6F_5 + F\bullet \rightarrow \sim C_6F_6\bullet \quad (7)$$

$$R\bullet + \sim C_6F_5 \rightarrow RC_6F_5\bullet, \ [R\bullet = \sim C_6F_4\bullet, \ \sim C_6F_5\bullet, \ \sim C_6F_5C_6F_4\bullet] \quad (8)$$

$$R\bullet + R\bullet \rightarrow RR \quad (9)$$

$$R\bullet + F\bullet \rightarrow RF \quad (10)$$

$$F\bullet + F\bullet \rightarrow F_2 \quad (11)$$

The simulation results were reasonable especially calculating the scission ratio, and predicting the radiation resistance of aromatic fluorocarbon. In the simulation, we assumed that the absorbed dose of polystyrene and fluorinated polystyrene relatively increased cascade area 1.6, 2.3, 2.8, and 3.2 Å. In this assumption, we converted the whole energy of γ-ray to the kinetic energy of atoms which were participated in the cascade. We expected that the experiment results were different from the simulation results because of the difference between the average molecular weight of atoms (experiment results) and atom number (simulation value). However, we coupled the simulation results with the scission rate and differential ratio of Mn (the number of molecular weight) and figured out that the trend of scission rate was similar in the results comparing the simulation data and experimental data. In this study, the scission process was only considered to predict the effect of fluorine. We will update this method with the crosslinking process to overcome the unforeseeable circumstances—prediction of radiation enhancement before the irradiation.

6. Conclusions

Generally, the high energy of the gamma-rays and the formation of radicals led to the chain scissioning, resulting in a decrease in molecular weight of the polymer. The produced alkyl free radicals react with oxygen to form peroxyl free radicals through the abstraction of hydrogen led to the cleavage of the C-H, C-C, and C-F chains.

When aromatic fluorinated polymers are irradiated, the observed yield of breaks obtained by splitting C-F and C-C bonds is often lower than those from the C-H and C-C bonds in aliphatic hydrocarbon polymers. Namely, the fluorine atoms are not readily abstracted from fluorinated polymer by general molecules and radicals such as H, CH_3, CF_3, and C_2F_5. This is because of radiation resistance due to the combination of C-F binding force and aromatic resonance structure. Actually, the aromatic hydrocarbon biphenyl and terphenyl are reported to be sufficiently resistant to radiation, the fluorine atoms of aromatic fluorocarbon (i.e., 1F-PS and 5F-PS) are also expected to be resistant to radiation. The more fluorine content in the benzene, the highly ordered π-stacking of fluorinated benzene increases charge carrier mobilities, suggesting stabilization against bond rupture by the radiation. So, it seems reasonable to conclude that the fluorine will be hardly abstracted from 5-F rather than 1-F based on simulation and measurement results [28].

Author Contributions: Conceptualization, Y.-H.Y., J.-H.P. and H.-J.G.; methodology, Y.-H.Y., H.-E.S. and N.-H.L.; software, Y.-H.Y., J.-H.P., M.-S.C., J.-H.M. and J.-Y.P.; validation, M.-S.C., J.-H.M. and J.-H.L.; formal analysis, Y.-H.Y., J.-H.L. and H.-E.S.; investigation, Y.-H.Y. and H.-E.S.; resources, Y.-H.Y. and H.-E.S.; data curation, Y.-H.Y. and H.-E.S.; writing—original draft preparation, Y.-H.Y. and H.-E.S.; writing—review and editing, Y.-H.Y., H.-E.S. and H.-J.G.; visualization, Y.-H.Y., H.-E.S. and H.-J.G.; supervision, H.-J.G. and N.-H.L.; project administration, N.-H.L.; funding acquisition, N.-H.L. All authors have read and agreed to the published version of the manuscript.

Funding: This work was supported by the Nuclear R&D program of the Ministry of Science and ICT.

Institutional Review Board Statement: Not applicable.

Informed Consent Statement: Not applicable.

Data Availability Statement: Not applicable.

Conflicts of Interest: The authors declare no conflict of interest.

References

1. Zhu, D.; Austerlitz, C.; Benhabib, S.; Mota, H.; Allison, R.R.; Campos, D. Study of a spherical phantom for Gamma knife dosimetry. *J. Appl. Clin. Med. Phys.* **2010**, *11*, 222–229. [CrossRef]
2. Gould, R.F. *Irradiation of Polymers, Copyright, Advances in Chemistry Series, FOREWORD*; American Chemical Society: Washington, DC, USA, 1967; pp. i–iv.
3. Onyiriuka, E.C.; Hersh, L.S.; Hertl, W. Surface modification of polystyrene by gamma-radiation. *Appl. Spectrosc.* **1990**, *44*, 808–811. [CrossRef]
4. Onyiriuka, E.C. The effects of high-energy radiation on the surface chemistry of polystyrene: A mechanistic study. *J. Appl. Polym. Sci.* **1993**, *47*, 2187–2194. [CrossRef]
5. Florin, R.E.; Wall, L.A. Gamma irradiation of fluorocarbon polymers. *J. Res. Natl. Bur. Stand. Sect. A Phys. Chem.* **1961**, *65*, 375. [CrossRef]
6. Carswell-Pomerantz, T.; Babanalbandi, A.; Dong, L.; Hill, D.J.; Perera, M.C.; Pomery, P.J.; Saadat, G.; Whittaker, A.K. *Stability and Stabilization of Polymers under Irradiation*; IAEA: Vienna, Austria, 1999; p. 111.
7. Von White, G., II; Tandon, R.; Serna, L.M.; Celina, M.C.; Bernstein, R. An Overview of Basic Radiation Effects on Polymers and Glasses. 2013. Available online: https://www.osti.gov/servlets/purl/1671997 (accessed on 21 November 2021).
8. Fuse, N.; Homma, H.; Okamoto, T. Remaining issues of the degradation models of polymeric insulation used in nuclear power plant safety cables. *IEEE Trans. Dielectr. Electr. Insul.* **2014**, *21*, 571–581. [CrossRef]
9. Gillen, K.T.; Clough, R.L. Predictive aging results in radiation environments. *Radiat. Phys. Chem.* **1993**, *41*, 803–815. [CrossRef]
10. Manas, D.; Ovsik, M.; Mizera, A.; Manas, M.; Hylova, L.; Bednarik, M.; Stanek, M. The effect of irradiation on mechanical and thermal properties of selected types of polymers. *Polymers* **2018**, *10*, 158. [CrossRef] [PubMed]
11. Polvi, J.; Luukkonen, P.; Nordlund, K.; Järvi, T.T.; Kemper, T.W.; Sinnott, S.B. Primary radiation defect production in polyethylene and cellulose. *J. Phys. Chem. B* **2012**, *116*, 13932–13938. [CrossRef]
12. Bowmer, T.N.; Cowen, L.K.; O'donnell, J.H.; Winzor, D.J. Degradation of polystyrene by gamma irradiation: Effect of air on the radiation-induced changes in mechanical and molecular properties. *J. Appl. Polym. Sci.* **1979**, *24*, 425–439. [CrossRef]
13. Shimizu, Y.; Mitsui, H. γ-radiation-induced crosslinking of polystyrene. *J. Polym. Sci. Polym. Chem. Ed.* **1979**, *17*, 2307–2316. [CrossRef]
14. Wall, L.A.; Brown, D.W. Gamma irradiation of polymethyl methacrylate and polystyrene. *J. Phys. Chem.* **1957**, *61*, 129–136. [CrossRef]
15. Wackerly, J.W.; Dunne, J.F. Synthesis of polystyrene and molecular weight determination by 1H NMR end-group analysis. *J. Chem. Educ.* **2017**, *94*, 1790–1793. [CrossRef]
16. Hanwell, M.D.; Curtis, D.E.; Lonie, D.C.; Vandermeersch, T.; Zurek, E.; Hutchison, G.R. Avogadro: An advanced semantic chemical editor, visualization, and analysis platform. *J. Cheminform.* **2012**, *4*, 17. [CrossRef]
17. Plimpton, S. Fast Parallel Algorithms for Short-Range Molecular Dynamics. *J. Comp. Phys.* **1995**, *117*, 1–19. [CrossRef]
18. Brenner, D.W. Empirical potential for hydrocarbons for use in simulating the chemical vapor deposition of diamond films. *Phys. Rev. B* **1990**, *42*, 9458. [CrossRef]
19. Van Duin, A.C.; Dasgupta, S.; Lorant, F.; Goddard, W.A. ReaxFF: A reactive force field for hydrocarbons. *J. Phys. Chem. A* **2001**, *105*, 9396–9409. [CrossRef]
20. Rahnamoun, A.; Van Duin, A.C.T. Reactive molecular dynamics simulation on the disintegration of Kapton, POSS polyimide, amorphous silica, and teflon during atomic oxygen impact using the ReaxFF reactive force-field method. *J. Phys. Chem. A* **2014**, *118*, 2780–2787. [CrossRef]
21. Beardmore, K.; Smith, R. Ion bombardment of polyethylene. *Nucl. Instrum. Meth. B* **1995**, *102*, 223–227. [CrossRef]
22. Nordlund, K.; Ghaly, M.; Averback, R.S.; Caturla, M.; de La Rubia, T.D.; Tarus, J. Defect production in collision cascades in elemental semiconductors and fcc metals. *Phys. Rev. B* **1998**, *57*, 7556. [CrossRef]
23. Robinson, M.T. Basic physics of radiation damage production. *J. Nucl. Mater.* **1994**, *216*, 1–28. [CrossRef]
24. Polvi, J.; Nordlund, K. Irradiation effects in high-density polyethylene. *Nucl. Instrum. Methods Phys. Res. Sect. B Beam Interact. Mater. At.* **2013**, *312*, 54–59. [CrossRef]
25. Oshima, A.; Horiuchi, H.; Nakamura, A.; Kobayashi, S.; Terui, A.; Mino, A.; Shimura, R.; Washio, M. Trapped radical behavior of electron beam irradiated polytetrafluoroethylene fine powder at various temperatures. *Sci. Rep.* **2021**, *11*, 10907. [CrossRef] [PubMed]
26. Yoshida, T.; Florin, R.E.; Wall, L.A. Stress relaxation of γ-irradiated fluorocarbon elastomers. *J. Polym. Sci. Part A Gen. Pap.* **1965**, *3*, 1685–1712. [CrossRef]

27. Huang, R.Y.M.; Westlake, J.F.; Sharma, S.C. Molecular weight distribution in radiation-induced polymerization. I. γ-radiation-induced free-radical polymerization of liquid styrene. *J. Polym. Sci. Part A-1 Polym. Chem.* **1969**, *7*, 1729–1747. [CrossRef]
28. Berger, R.; Resnati, G.; Metrangolo, P.; Weber, E.; Hulliger, J. Organic fluorine compounds: A great opportunity for enhanced materials properties. *Chem. Soc. Rev.* **2011**, *40*, 3496–3508. [CrossRef] [PubMed]

materials

MDPI

Article

Preliminary Study on the Simulation of a Radiation Damage Analysis of Biodegradable Polymers

Ha-Eun Shim †, Yeong-Heum Yeon †, Dae-Hee Lim, You-Ree Nam, Jin-Hyung Park, Nam-Ho Lee and Hui-Jeong Gwon *

Advanced Radiation Technology Institute, Korea Atomic Energy Research Institute, 29 Geumgu-gil, Jeongeup 56212, Korea; she0805@kaeri.re.kr (H.-E.S.); yhyeon@kaeri.re.kr (Y.-H.Y.); daehee@kaeri.re.kr (D.-H.L.); yrnam@kaeri.re.kr (Y.-R.N.); jhpak@kaeri.re.kr (J.-H.P.); nhlee@kaeri.re.kr (N.-H.L.)

* Correspondence: hjgwon@kaeri.re.kr; Tel.: +82-63-570-3087

† These authors equally contributed to this work.

Abstract: In this study, biodegradable poly(L-lactide-co-ε-caprolactone) (PLCL) and poly(L-co-D,L lactide) (PLDLA) were evaluated using Geant4 (G4EmStandardPhysics_option4) for damage simulation, in order to predict the safety of these biodegradable polymers against gamma ray sterilization. In the PLCL damage model, both chain scission and crosslinking reactions appear to occur at a radiation dose in the range 0–200 kGy, but the chain cleavage reaction is expected to be relatively dominant at high irradiation doses above 500 kGy. On the other hand, the PLDLA damage model predicted that the chain cleavage reaction would prevail at the total irradiation dose (25–500 kGy). To verify the simulation results, the physicochemical changes in the irradiated PLCL and PLDLA films were characterized by GPC (gel permeation chromatography), ATR-FTIR (attenuated total reflection Fourier transform infrared), and DSC (difference scanning calorimetry) analyses. The Geant4 simulation curve for the radiation-induced damage to the molecular weight was consistent with the experimentally obtained results. These results imply that the pre-simulation study can be useful for predicting the optimal irradiation dose and ensuring material safety, particularly for implanted biodegradable materials in radiation processing.

Keywords: PLCL; PLDLA; chain scission; degradation; gamma-ray; Geant4; simulation; radiation damage model

Citation: Shim, H.-E.; Yeon, Y.-H.; Lim, D.-H.; Nam, Y.-R.; Park, J.-H.; Lee, N.-H.; Gwon, H.-J. Preliminary Study on the Simulation of a Radiation Damage Analysis of Biodegradable Polymers. *Materials* **2021**, *14*, 6777. https://doi.org/10.3390/ma14226777

Academic Editor: Vladimir Krsjak

Received: 7 October 2021
Accepted: 5 November 2021
Published: 10 November 2021

1. Introduction

Nowdays, biodegradable polymers are widely used as materials for medical devices [1]. Biodegradable polymers, used for decades, include polyesters and their copolymers, such as poly(L-lactic acid) (PLA), poly(ε-caprolactone) (PCL), poly(L-lactide-co-ε-caprolactone) (PLCL), and poly(L-co-D,L lactide) (PLDLA). Among the various materials, PLCL and PLDLA are very valuable materials used in medical applications as implantable devices because of their excellent flexibility and biodegradability [2–4]. Sterilization is essential for implantable devices [5], and some established sterilization methods include dry heat, ethylene oxide, steam, and radiation methods [6]. In particular, the gamma or electron beam sterilization process is performed at room temperature and has the advantage of a short sterilization time and low risk of toxic residues [7]. In addition, it has a high sterilization effect for substances that struggle to penetrate into other sterilizing agents [8]. Packaging is used to protect the bioimplantable device from moisture and ions inside the human body [9], the material is sterilized in the entire volume of the product together with the packaging. Because of these advantages, gamma irradiation is the most commonly used method for the sterilization of materials with a high transmittance [10].

However, free radicals generated by radiation energy can propagate within the polymer chain structure and cause a chain reaction, leading to crosslinking [11–13]. Therefore, aliphatic polyesters are decomposed in the radiation sterilization process, and the decomposition temperature may change depending on the polymer composition [7,14,15]. In general, the effect of radiation on polymers can cause changes in various properties, such as chemical composition, crystallinity, molecular weight and density, depending on the radiation dose, dose rate, and temperature [6]. In particular, because biodegradable, polymer-based, implantable devices are very important for the performance and service life of materials [16,17], it is necessary to analyze the changes in the properties of polymer materials according to irradiation dose, in preparation for the possibility of material decomposition during the radiation sterilization process. In addition, radiation sterilization is the most suitable and useful material for human insertion among the currently available sterilization methods. Therefore, it is necessary to secure the optimal sterilization dose for each polymer material according to its intended use.

In contrast, irradiation is well known as a very convenient tool for sterilizing biodegradable polymers and transforming polymer materials through crosslinking, grafting, and decomposition [18]. Because crosslinking using radiation may generate radicals within the polymer chain, the functionalization process may be omitted, and the crosslinking reaction may be performed at a low temperature due to its excellent penetrating power. In particular, irradiation with a high-energy radiation such as gamma rays and electron beams was used for the processing and crosslinking of polymers [19,20].

However, radiation accompanies the chain scission reaction and crosslinking simultaneously and may affect basic properties such as reducing the glass transition, crystallization and melting temperature of the polymer. Other physical properties, such as gas permeability, may differ from those of conventional polymers. In particular, because crosslinking generally reduces the degradability of polymers, it can negatively affect the properties of materials designed for degradable devices after the insertion into the human body [16]. Radiation treatment can control the biodegradation time [21]. A compromise must be found between the mechanical properties and the biodegradation time. Therefore, it is expected that the time and cost of material development can be dramatically shortened if the optimal radiation irradiation conditions that are suitable for the required properties can be selected through a pre-simulation study.

In addition, to analyze the damage to polymers caused by radiation, it is necessary to understand radiation physics, dosimetry, chemical analysis, and instrumental analysis applied to the modeling and simulation of the radiation environment [22]. Therefore, it is necessary to investigate the interaction of the polymer of interest with gamma rays. However, because it is difficult to rely solely on repeated testing and dose selection for an accurate analysis [23], some researchers conducted simulation-based correlation studies [24–26]. Ghosal et al. performed theoretical simulations to determine the correlation between the morphological changes due to gamma irradiation and other properties such as molecular weight distribution, intrinsic viscosity, and ionic conductivity [24]. Saha et al. conducted a computer simulation study to examine the effects of gamma irradiation on properties such as molecular weight distribution and viscosity [25]. In particular, the number average molecular weight can be an indicator of the critical dose that causes changes in the physical properties of polymers due to radiation [27]. In addition, the quantitative analysis and interpretation of the number average molecular weight is important for understanding the decomposition behavior of polymer materials according to the radiation dose within the simulation.

In this study, the average molecular weight of the polymer was selected as a parameter to conduct a basic simulation of radiation damage analysis for biodegradable polymers (PLCL and PLDLA). Figure 1 shows the chemical structures of PLCL and PLDLA. In preparation for the possibility of material decomposition during radiation (gamma-ray) sterilization, changes in the characteristics of PLCL and PLDLA according to irradiation dose were observed. In addition, a Geant4 simulation was performed to predict the structural damage dose range in the gamma ray sterilization and processing by selecting this as a damage model, and a comparative verification study was conducted to verify the change in the average molecular weight according to the irradiation dose.

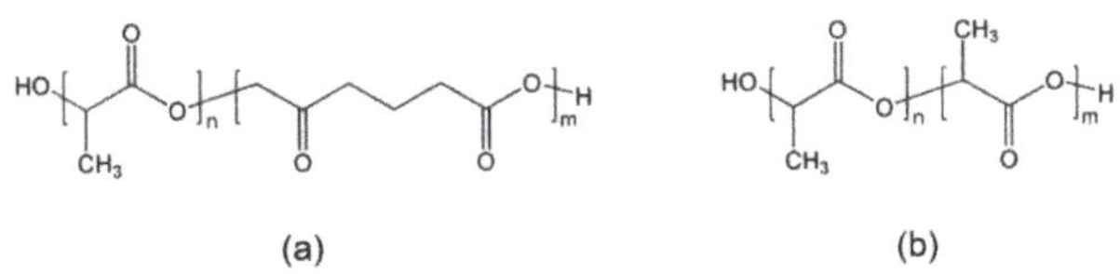

Figure 1. Chemical structure of polymer (**a**) PLCL and (**b**) PLDLA.

2. Materials and Methods

2.1. Modeling and Simulation

2.1.1. Geant4 Simulation for Gamma Ray Fluence Calculation

To calculate the radiation dose of polymer films, Geant4 (version 10.6p02, CERN, Conseil Européenne pour la Recherche Nucléaire, Meyrin, Switzerland) was used to perform the simulation. First, the radiation dose of the high-level gamma radiation device was simulated. Using the "GeneralParticleSource (GPS)" and "G4RadioactiveDecayPhysics" modules in Geant4, the ^{60}Co decay scheme was modeled. The energy absorbed by an alanine dosimeter (cylindrical type with a radius of 0.24 cm and height of 0.3 cm), placed 15.8 cm from the ^{60}Co source, was calculated. The "G4EmStandardPhysics_option4" physics model was used to calculate the absorbed dose of the alanine (material: $C_3H_7NO_2$, density: 1.42 g/cm^3) dosimeter when exposed to 1173 keV and 1332 keV gamma rays simultaneously released from the ^{60}Co source. The dose amount necessary to calculate the same radiation dose as experimental conditions at the target position was converted to fluence.

2.1.2. Geant4 Simulation for Absorbed Dose Calculation

Simulations were performed to calculate the absorbed dose per unit mass (D, eV/g) after irradiating the PLCL and PLDLA films with the previously calculated gamma ray fluence. The polymer film was the same size as the actual film, measuring $2 \times 2 \times 0.1$ cm^3. The PLCL film was made of $HO[C_3H_4O_2]_n[C_6H_{10}O_2]_mCH_3$, (n:m = 70:30), and had a density of 1.2 g/mL. The PLDLA film was made of $HO[C_3H_4O_2]_n[C_3H_4O_2]_mCH_3$, (n:m = 70:30), and had a density of 1.2 g/mL. The PLCL and PLDLA films were placed 15.8 cm away from the ^{60}Co source, and the absorbed dose per unit mass was calculated as summarized in Table 1.

Table 1. Absorbed dose per unit mass of PLCL and PLDLA models according to gamma radiation dose.

Dose (kGy)	PLCL	PLDLA
	Absorbed Dose per Unit Mass (eV/g)	
25	1.47×10^{21}	2.73×10^{20}
50	2.95×10^{21}	5.47×10^{20}
100	5.90×10^{21}	1.09×10^{21}
200	1.18×10^{22}	2.19×10^{21}
500	2.95×10^{22}	5.47×10^{21}

2.1.3. Prediction of Damage to Polymer Materials Using a Radiation Damage Model

The relationship between the radiation dose and number average molecular weight can be derived as follows: The number average molecular weight ($M_{n,0}$, g/mole) for the polymer sample is as follows [28]:

$$M_{n,0}\ (\text{g/mole}) = wN_A/N_0 \tag{1}$$

where w denotes the weight (g) of the polymer sample, N_A is Avogadro's number, and N_0 is the total number of molecules (initial molecules) in w before irradiation. From Equation (1), N_0 can be rearranged as follows:

$$N_0 = wN_A/M_n \tag{2}$$

If the dose is expressed as D (eV/g), we can calculate Dw, that is, the total absorbed dose by the sample. Therefore, the number of newly formed molecules (N^*) inside the polymer due to irradiation is obtained as follows:

$$N^* = KDw \tag{3}$$

where K denotes the polymer structure constant that represents the resistance to radiation. It can be replaced by G, which is defined as the number of molecules or atoms produced per 100 eV of energy.

The G-value is generally expressed as G_s for the number of scissions due to exposure, or G_x for the number of crosslinking reactions. The production of new molecules in relation to the number of scissions ($N_s{}^*$) and crosslinking ($N_x{}^*$) can be described as follows:

$$N_s{}^* = (G_s/100)\ Dw \tag{4}$$

$$N_x{}^* = (G_x/100)\ Dw \tag{5}$$

From the perspective of the number of molecules in the polymer sample, chain scissions increase the number of molecules, whereas crosslinking decreases the number of molecules. Therefore, the number average molecular weight ($M_n{}^*$) of the polymer when the scissions and crosslinking reactions occur competitively with the absorbed dose D is obtained as follows:

$$M_n{}^* = wN_A/(N_0 + \alpha N_s{}^* - \beta N_x{}^*) \tag{6}$$

where the total mass of the polymer is assumed to be constant during exposure, α and β denote constants used to consider the change in the initial number of molecules (N_0) when chain scission and crosslinking events occur; 0.5 and 2 are applied, respectively.

Based on Equations (4)–(6), the polymer damage model that considers radiation-induced chain scission and crosslinking is as follows:

$$1/M_n{}^* = 1/M_{n,0} + [(\alpha G_s - \beta G_x)/100N_A]\ D \tag{7}$$

where $M_{n,0}$ and G-value (G_s and G_x) were obtained from the experimental results. The radiation damage model of Equation (7) was used to simulate a biodegradable polymer film.

2.2. *Film Preparation and Radiation Measurement for Simulation Verification*

2.2.1. Materials

Poly(L-lactide-co-ε-caprolactone) (PLCL, 70/30) and poly(L-lactide-co-D,L-Lactide) (PLDLA 70/30) were commercially purchased from RESOMER® (Evonik Health Care Evonik Industries AG, Essen, Germany). Chloroform ($CHCl_3$) was selected as a solvent and was supplied by Showa, Tokyo, Japan. All other reagents and solvents were of analytical grade and were used without further purification.

2.2.2. Sample Irradiation

PLCL and PLDLA films were prepared by a solution casting method using $CHCl_3$ as a solvent. The PLCL and PLDLA powders were dissolved in $CHCl_3$ to obtain 10 wt% and 4 wt% polymer solutions, respectively. The PLCL and PLDLA films were fabricated by pouring a polymer solution into a well-cleaned glass plate and evaporating the solvent in air at room temperature. Dried films were peeled off manually from the glass plate and dried in vacuum oven for 24 h at room temperature. Prior to gamma irradiation, the prepared films were packed with nitrogen gas in glass vials. Gamma irradiation was thereafter performed using a gamma ^{60}Co source on the samples with different radiation doses (25, 50, 100, 200, and 500 kGy) at a dose rate of 10 kGy/h. The ^{60}Co source (MDS Nordion, Ottawa, Canada, IR 221 n wet storage type C-188) was located at the Korea Atomic Energy Research Institute (KAERI), Jeongeup, Republic of Korea.

2.2.3. Attenuated Total Reflection Fourier Transform Infrared Spectroscopy (ATR-FTIR)

ATR spectra of gamma-ray-irradiated biodegradable polymer films were obtained using an ATR-FTIR spectrophotometer (Bruker TENSOR 37, Bruker Corporation, Billerica, MA, USA). The spectra were measured in the wavenumber range from 500 to 4000 cm^{-1} in ATR mode. Spectra were recorded using Bruker OPUS software (version 8.5, Bruker Corporation, Billerica, MA, USA) at a resolution of 4 cm^{-1}.

2.2.4. Thermal Analysis

The thermal properties of the polymer films were tested by difference scanning calorimetry (DSC, Q100, TA Instruments, New Castle, DE, USA). The DSC thermograms of the polymer films were measured from −50 to 250 °C under a nitrogen atmosphere at a heating rate of 5 °C/min, and then were cooled at room temperature. A second heating cycle was then observed by heating the samples from −50 to 250 °C.

2.2.5. Average Molecular Weight

Changes in the molecular weights of the irradiated film were determined by gel permeation chromatography (GPC, PL-GPC 110, Polymer Laboratories, Church Stretton, UK) with the eluent of $CHCl_3$ at a flow rate of 1.0 mL/min at 40 °C. The GPC system was equipped with columns of PLgel Guard column 5 μm, PLgel 10 μm Mixed B and PLgel 5 μm 10,000 A (Polymer Laboratories, Church Stretton, UK) calibrated with polystyrene standards.

The changes in molecular weight are related to the radiation chemical yields of crosslinking (G_x) and chain scission (G_s), which determines the extent of chain scission or crosslinking during gamma-ray irradiation, and can be calculated from the following equations [27]:

$$1/M_w = 1/M_{w,0} + (G_s/2 - 2G_x)D \times 1.038 \times 10^{-6} \quad (8)$$

$$1/M_n = 1/M_{n,0} + (G_s - G_x)D \times 1.038 \times 10^{-6} \quad (9)$$

where $M_{w,0}$ and $M_{n,0}$ are the weight and number average molecular weight of unirradiated films. M_w and M_n are the corresponding values following exposure to irradiation dose, D. A ratio of G_s/G_x greater than 4 would indicate that chain scission is more prominent for nitrogen atmospheres [29].

3. Results and Discussion

3.1. Modeling and Simulation

Figure 2 shows the number average molecular weight change according to the gamma irradiation dose of the radiation damage model PLCL and PLDLA films through Geant4 simulation. The blue triangle represents the simulation result considering both chain scission and crosslinking in the radiation damage model, whereas the red square represents the simulation result considering only the chain scission in the radiation damage model. The $M_{n,0}$ and G-values (G_s and G_x) of PLCL and PLDLA obtained from the experimental results were used in Equation (7) for the radiation damage model of the polymer considering the radiation cleavage and crosslinking derived in this study, respectively.

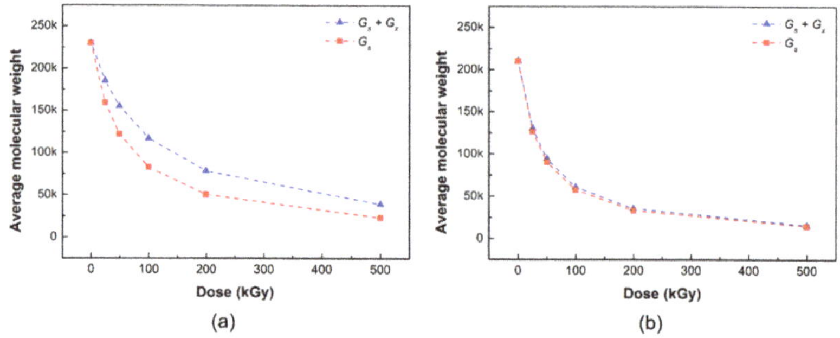

Figure 2. Number average molecular weight of (**a**) PLCL and (**b**) PLDLA radiation damage models depending on the gamma-ray irradiation in simulation (blue triangles: simulation of chain scission and crosslinking; red squares: simulation of chain scission).

In the simulation results, the average molecular weight of the radiation damage model PLCL decreased with the increasing irradiation dose (Figure 2a). In the two damage models, the difference in the number average molecular weight in the range 25–200 kGy of the irradiation dose appeared to gradually widen, but the gap seemed to narrow at the irradiation dose of 500 kGy; it appeared that both the chain cleavage and crosslinking reactions occurred up to the range before the irradiation dose of 200 kGy, but the chain cleavage reaction was expected to be relatively dominant at a high irradiation dose of 500 kGy or more.

Figure 2b shows the number average molecular weight changes, according to the gamma radiation dose of the radiation damage model PLDLA film, through the Geant4 simulation. In the simulation results, the number average molecular weight of the radiation damage model PLDLA decreased with the radiation dose. Interestingly, when only chain cleavage was considered in the radiation damage model, there was no significant difference from when both chain cleavage and crosslinking were considered, that is, in the PLDLA damage model, it can be inferred that the chain cleavage reaction prevails over crosslinking.

3.2. *Radiation Assessment for Simulation Verification*

3.2.1. ATR-FTIR Spectroscopy

Figure 3 shows the ATR-FTIR spectra of PLCL and PLDLA before and after gamma-ray irradiation. The ATR-FTIR spectrum of the PLCL film was observed from 1300 to 1000 cm^{-1}, which is related to the stretching of C−O bonds in the ester, found in the long alkyl chain of the polymer structure [30,31]. The C−O and C−O−C groups exhibited stretching peaks at 1188, 1037, and 1079 cm^{-1} for the PLCL film both before and after exposure. The stretching of −C=O (carbonyl) appeared as an intense peak at 1750 cm^{-1}. In addition, the asymmetric stretching vibration of the $-CH_3$ and $-CH_2$ groups were observed at 2992 and 2943 cm^{-1} for the PLCL film before and after exposure [29,31,32]. Another bond related to the symmetric vibration of $-CH_2$ was observed at 2872 cm^{-1} [29]. Bonds associated with the asymmetrical and symmetrical stretching of the $-CH_3$ group exhibited peaks at 1450 and 1361 cm^{-1}, respectively [33]. Small peaks were also observed for the C−H bending vibrations at 757 and 864 cm^{-1} [34].

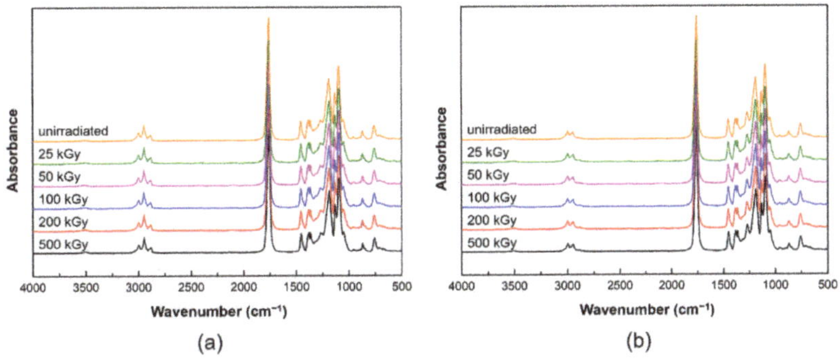

Figure 3. ATR-FTIR spectra before and after gamma-ray irradiation: (**a**) PLCL and (**b**) PLDLA polymer films.

Figure 3b shows the ATR-FTIR spectra of PLDLA before and after gamma-ray irradiation. The C−O and C−O−C groups exhibited stretching peaks at 1182 and 1081 cm^{-1} for the PLDLA film both before and after exposure. The stretching of −C=O (carbonyl) appeared as an intense peak at 1746 cm^{-1}. In addition, the asymmetric stretching of −CH was observed at 2994 and 2945 cm^{-1} for the PLDLA film before and after exposure. Another bond related to the bending vibration of $-CH_3$ was observed at 1452 cm^{-1} [33]. Small peaks were also observed for the C−H bending vibrations at 759 and 872 cm^{-1} [34].

The PLCL and PLDLA films did not display significant differences before and after exposure due to minor radiation-induced chemical changes occurring in the polymer chain [35,36]. This implies that there was no change in the functional group inside the polymer after irradiation, and this result indicated that the possibility of the creation of a new bond, which was one of the parts to be considered in the decomposition mechanism of the polymer, could be excluded.

3.2.2. Thermal Analysis

Figure 4 shows the DSC curves of PLCL and PLDLA before and after gamma-ray irradiation. The DSC curve of the PLCL film before exposure had a melting point (T_m) of 159.78 °C. No significant differences were observed up to a dose of 100 kGy. However, T_m decreased rapidly when the dose was increased to 200 and 500 kGy, indicating that gamma-ray irradiation resulted in the degradation and scissioning of the main chain [20,37].

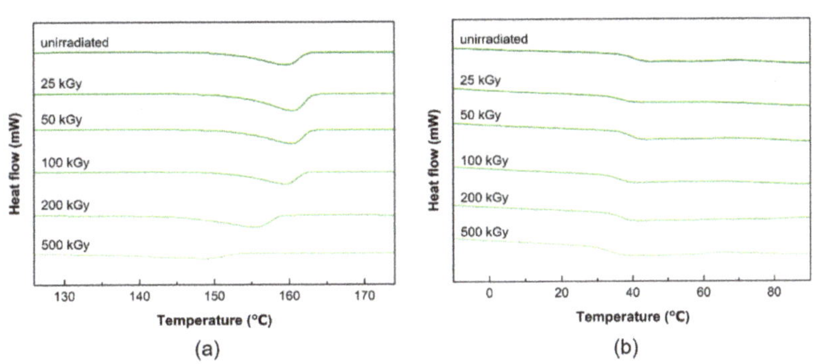

Figure 4. DSC thermograms of the before and after gamma-ray irradiation (**a**) PLCL and (**b**) PLDLA polymer films.

In the DSC curve of the PLDLA film before irradiation, a glass transition (T_g) was observed at 41.75 °C. After gamma irradiation, there was no significant change at 100 kGy or less, and a decrease in T_g was observed at 200 kGy. The T_g of a polymer was related to its molecular weight, and T_g decreases as the average molecular weight decreases [38]. Consequently, it was confirmed that the temperature at which thermal transition occurs changed when the dose was greater than 100 kGy.

3.2.3. Average Molecular Weight

The changes in the average molecular weights (M_w and M_n) of the PLCL and PLDLA films before and after gamma-ray irradiation are shown in Figure 5. The oxidation of polymers during exposure reduces crosslinking, increases degradation, or causes chain scissions [24]. Thus, the formation of radicals after exposure results in chain scissions, which in turn lowers the molecular weight. Chain scissions usually occur when polymers are in the amorphous phase [25].

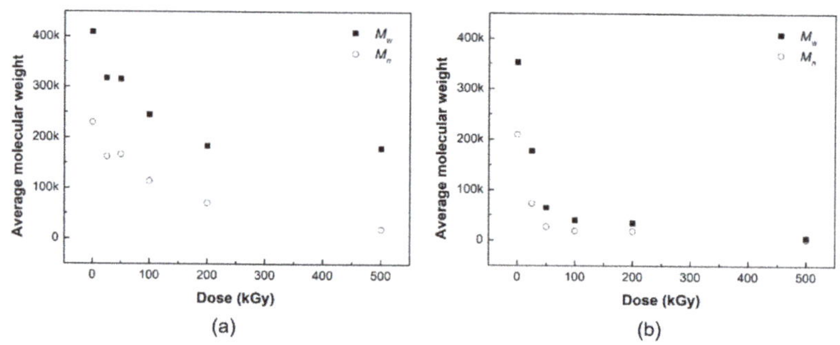

Figure 5. Average molecular weight (black squares: M_w, white circles: M_n) of (**a**) PLCL and (**b**) PLDLA depending on the gamma-ray irradiation.

At higher doses of at least 200 kGy, chain scission occurs because of the alkyl free radicals that react with oxygen to form peroxyl free radicals through hydrogen abstraction [26,27]. This type of chain scission has no significant effect on the decrease in the average molecular weight, given the relatively higher increase in chain scission, compared to crosslinking events under higher doses [39]. The number of alkyl free radicals was greater than that of peroxyl free radicals under higher radiation doses because oxygen was not present under our experimental conditions. Alkyl free radicals had less influence on chain scissions than peroxyl free radicals, and were more likely to undergo rebonding or crosslinking in crystalline and amorphous segments [40].

Hydrogen abstraction due to chain-breaking radicals in the weakest C–H bonds contributes to an increase in alkyl radicals or $-C(CH_3)-$ radicals. The average molecular weight decreased with the degradation of the polymer during exposure [41]. From Table 2, the high G-value (G_s/G_x = 8.7) exhibits a degree of chain scission in the irradiated PLCL film after exposure. The decrease in molecular weight at low radiation doses was due to chain scission by alkyl free radicals [26]. High-energy irradiation forms radicals played a role in the degradation of polymers [41]. In other words, the chain appears to be affected at a dose of 30 kGy or higher, but exhibits a more distinct difference in molecular weight at higher doses of at least 200 kGy. Even in the absence of oxygen in the air, radicals that cause chain cleavage may be formed by radiation energy, which may affect the average molecular weight of the polymer.

Table 2. Chain scission (G_s) and crosslinking (G_x) radiation yields of PLCL and PLDLA irradiated.

Polymer (dose)	G_s	G_x	G_s/G_x
PLCL (200)	0.158	0.018	8.7
PLDLA (100)	1.394	0.026	52.7

Polymers with oxygen atoms are known to exhibit a very high sensitivity to radiation [2], and similar studies were conducted on structures containing oxygen in the polymer backbone [29]. Methine groups in the polymer backbone appear to be important in the radiolysis of biodegradable polymers containing oxygen atoms. Previously, Nugroho et al. reported a study on the gamma-ray degradation of PLA [42]. This polymer contains an ester linkage and a cleavage site in the methine group. The cleavage of the ester bond at this cleavage site causes crosslinking with a relatively low yield, whereas cleavage at the methine group causes the chain scission of the polymer [17,42].

In general, the radicals generated at the ends of polymer chains generate new radicals, move to adjacent polymer chains, or cause hydrogen abstraction [42]. In addition, double bonds are formed at the chain ends after hydrogen abstraction. In PLDLA, a copolymer comprising of isomers of PLA, and both crosslinking reactions, can cause decomposition reactions due to cleavage of ester bonds in the polymer and hydrogen abstraction of the methine groups [20]. In this study, when interpreted only as a result of the decrease in the average molecular weight of PLDLA according to gamma radiation dose, the radicals generated inside the polymer react well with each other and the decomposition reaction is relatively dominant; therefore, it is believed that primarily chain cleavage occurs. Consequently, it is observed that the polymer chain is affected by radiation even at a low gamma irradiation of 25 kGy or higher, but it can be confirmed that a distinct change is exhibited with a molecular weight at 50 kGy or higher.

3.2.4. Modeling and Simulation Verification

Figure 6 shows the results of the simulation of the radiation damage model and the change in the number average molecular weight of the PLCL and PLDLA films according to the gamma radiation dose. The blue triangle represents the simulation result considering both chain scission and crosslinking in the radiation damage model, the red square represents the simulation result considering only the chain scission in the radiation damage model, and the white circle represents the experimental result obtained from the radiation measurement evaluation.

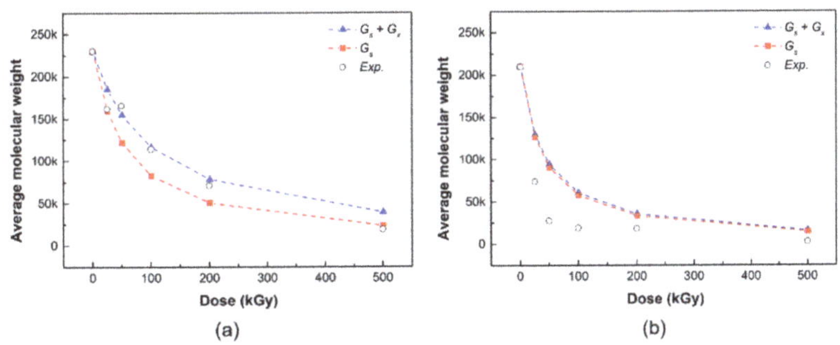

Figure 6. Comparison of number average molecular weight of (**a**) PLCL and (**b**) PLDLA in simulation and measurement evaluation (blue triangles: simulation of chain scission and crosslinking, red squares: simulation of chain scission, white circles: experimental value).

The simulation results predicted that, in the radiation damage model, both chain cleavage and crosslinking reactions occurred at low doses before 200 kGy, but chain cleavage reactions were relatively dominant at high doses of 500 kGy or higher (Figure 6a). The number average molecular weight of the PLCL film decreased according to the gamma radiation dose in the radiation measurement results, which exhibited a similar trend to the Geant4 simulation results. The actual evaluation results were similar to those of the radiation damage model when both chain scission and crosslinking were considered at a radiation dose in the range 0–200 kGy. Meanwhile, the actual evaluation results at a dose of 500 kGy were similar to the simulation results of the radiation damage model considering only chain scission. Consequently, it can be determined that PLCL has a more dominant chain cleavage reaction than crosslinking at a high radiation dose of at least 500 kGy, and the simulation of the radiation damage model can be verified through a comparison with the experimental results.

The simulation results of PLDLA exhibited no significant difference when only chain cleavage was considered in the radiation damage model, compared to when both chain cleavage and crosslinking were considered (Figure 6b). Therefore, the PLDLA damage model predicted that the chain cleavage reaction would prevail at the total irradiation dose (25–500 kGy). The results of the radiation measurement demonstrated that the polymer chain was affected by radiation even at a low irradiation dose of 25 kGy or higher, and that the chain cleavage reaction was dominant, with a distinct molecular weight change at 50 kGy or higher. Consequently, unlike PLCL in which cleavage was predominant at high doses, in the case of PLDLA, it could be determined that the chain cleavage reaction was dominant at the total irradiation dose (25–500 kGy), and the simulation could be verified through a comparison with experimental results.

In addition, in Equation (7) derived from this study, the slopes α and β are correlated with the G value, and thus the radiation damage model is an important factor in simulating the actual evaluation experiment. To reflect the change in the initial number of molecules (N_0) due to molecular formation through chain cleavage and crosslinking, 0.5 and 2 were applied to the slopes α and β, respectively. The average molecular weight change curve was also determined.

Regarding the degradation of biodegradable polymers, as well as the results of our study, the results of modeling and simulation studies by other researchers were also reported [43–45]. This work focused on the average molecular weight required to conduct a basic simulation of radiation damage analysis for biodegradable polymers. However, the analysis reported in this paper corresponds to a simplified model of the interaction of ionizing radiation through only single ionization. It was not fully considered that, as a result of radiolysis, apart from molecular hydrogen, molecular oxygen was also released [46]. Thus, it is necessary to carry out additional research considering post-radiation, oxidative degradation processes.

4. Conclusions

In this study, PLCL and PLDLA, which are representative, implantable polymer materials, were selected as target material models to conduct a predictive simulation study on the radiation damage and prediction of biodegradable polymer materials.

A decomposition evaluation using gamma rays was performed through the characterization (ATR-FTIR, GPC, and DSC thermal analysis) of PLCL and PLDLA films according to the irradiation dose. No clear structural decomposition was observed in the ATR-FTIR results, but changes in the melting point (T_m) and average molecular weight (M_n and M_w) of the PLCL film decreased as the irradiation dose increased, as shown by the DSC and GPC analysis results. Similarly, in the PLDLA film, no clear structural decomposition was observed in the ATR-FTIR results. Yet, in the GPC analysis results, a decrease in the average molecular weight (M_n and M_w) of the PLDLA film was observed as the irradiation dose increased. In contrast, a change in the glass transition temperature (T_g) was observed at irradiation doses greater than 100 kGy, and it seemed that the optimal sterilization dose to maintain thermal stability could be derived.

In addition, PLCL and PLDLA films were selected as damage analysis and prediction models and were manufactured in the form of films; a Geant4 simulation was performed using gamma rays. Consequently, a number average molecular weight (M_n) change curve similar to the actual measurement result was confirmed. This study is a preliminary study for the analysis and prediction of radiation damage of biodegradable polymer materials using gamma rays, which could be possible.

Author Contributions: Conceptualization, H.-E.S. and H.-J.G.; methodology, H.-E.S. and Y.-H.Y.; software, Y.-H.Y. and J.-H.P.; validation, D.-H.L. and Y.-R.N.; formal analysis, H.-E.S.; investigation, Y.-R.N.; data curation, J.-H.P. and N.-H.L.; writing—original draft preparation, H.-E.S. and Y.-H.Y.; writing—review and editing, H.-J.G.; supervision, N.-H.L.; project administration, H.-J.G. All authors have read and agreed to the published version of the manuscript.

Funding: This research was supported by the Nuclear R&D program of the Ministry of Science and ICT.

Institutional Review Board Statement: Not applicable.

Informed Consent Statement: Not applicable.

Data Availability Statement: Not applicable.

Conflicts of Interest: The authors declare no conflict of interest.

References

1. Zada, M.H.; Kumar, A.; Elmalak, O.; Mechrez, G.; Domb, A.J. Effect of Ethylene Oxide and Gamma (γ-) Sterilization on the Properties of a PLCL Polymer Material in Balloon Implants. *ACS Omega* **2019**, *4*, 21319–21326. [CrossRef] [PubMed]
2. Fernández, J.; Etxeberria, A.; Ugartemendia, J.M.; Petisco, S.; Sarasua, J.R. Effects of chain microstructures on mechanical behavior and aging of a poly(L-lactide-co-ε-caprolactone) biomedical thermoplastic-elastomer. *J. Mech. Behav. Biomed.* **2012**, *12*, 29–38. [CrossRef] [PubMed]
3. Kim, D.H.; Vaccaro, A.R.; Fessler, R.G. *Spinal Instrumentation*, 2nd ed.; Thieme Press: New York, NY, USA, 2018.
4. Kluger, P.; Wyrwa, R.; Weisser, J.; Maierle, J.; Votteler, M.; Rode, C.; Schnabelrauch, M.; Walles, H.; Schenke-Layland, K. Electrospun poly(d/l-lactide-co-l-lactide) hybrid matrix: A novel scaffold material for soft tissue engineering. *J. Mater. Sci. Mater. Med.* **2010**, *21*, 2665–2671. [CrossRef] [PubMed]
5. Bernkopf, M. Sterilisation of bioresorbable polymer implants. *Med. Device Technol.* **2007**, *26*, 28–29.
6. Tipnis, N.P.; Burgess, D.J. Sterilization of implantable polymer-based medical devices: A review. *Int. J. Pharm.* **2018**, *544*, 455–460. [CrossRef]
7. Jain, S.; Yassin, M.A.; Fuoco, T.; Mohamed-Ahmed, S.; Vindenes, H.; Mustafa, K.; Finne-Wistrand, A. Understanding of how the properties of medical grade lactide based copolymer scaffolds influence adipose tissue regeneration: Sterilization and a systematic in vitro assessment. *Mater. Sci. Eng.* **2021**, *124*, 112020. [CrossRef]
8. Rychter, P.; Śmigiel-Gac, N.; Pamuła, E.; Smola-Dmochowska, A.; Janeczek, H.; Prochwicz, W.; Dobrzyński, P. Influence of radiation sterilization on properties of biodegradable lactide/glycolide/trimethylene carbonate and lac-tide/glycolide/ε-caprolactone porous scaffolds with shape memory behavior. *Materials* **2016**, *9*, 64. [CrossRef]
9. Teo, A.; Mishra, A.; Park, I.; Kim, Y.J.; Park, W.-T.; Yoon, Y.-J. Polymeric Biomaterials for Medical Implants and Devices. *ACS Biomater. Sci. Eng.* **2016**, *2*, 454–472. [CrossRef]
10. Pérez Davila, S.; González Rodríguez, L.; Chiussi, S.; Serra, J.; González, P. How to Sterilize Polylactic Acid Based Medical Devices? *Polymers* **2021**, *13*, 2115. [CrossRef]
11. Hasanain, F.; Guenther, K.; Mullett, W.M.; Craven, E. Gamma sterilization of pharmaceuticals—A review of the irradiation of excipients, active pharmaceutical ingredients, and final drug product formulations. *PDA J. Pharm Sci. Technol.* **2014**, *68*, 113–137. [CrossRef]
12. Montanari, L.; Cilurzo, F.; Valvo, L.; Faucitano, A.; Buttafava, A.; Groppo, A.; Genta, I.; Conti, B. Gamma irradiation effects on stability of poly(lactide-co-glycolide) microspheres containing clonazepam. *J. Control. Release* **2001**, *75*, 317–330. [CrossRef]
13. Aquino, K.A.D.S. Sterilization by Gamma Irradiation. *Gamma Radiat.* **2012**, *9*, 172–202. [CrossRef]
14. Taraballi, F.; Corradetti, B.; Minardi, S.; Powel, S.; Cabrera, F.; Van Eps, J.L.; Weiner, B.K.; Tasciotti, E. Biomimetic colla-genous scaffold to tune inflammation by targeting macrophages. *J. Tissue Eng.* **2016**, *7*, 1–13. [CrossRef] [PubMed]
15. Odelius, K.; Plikk, P.; Albertsson, A.-C. The influence of composition of porous copolyester scaffolds on reactions induced by irradiation sterilization. *Biomaterials* **2008**, *29*, 129–140. [CrossRef]
16. Bednarek, M.; Borska, K.; Kubisa, P. Crosslinking of Polylactide by High Energy Irradiation and Photo-Curing. *Molecules* **2020**, *25*, 4919. [CrossRef]
17. Kumar, A.; Tumu, V.R.; Chowdhury, S.R.; Ramana Reddy, S.V.S. A green physical approach to compatibilize a bio-based poly (lactic acid)/lignin blend for better mechanical, thermal and degradation properties. *Int. J. Biol. Macromol.* **2018**, *121*, 588–600. [CrossRef]
18. Manas, D.; Ovsik, M.; Mizera, A.; Manas, M.; Hylova, L.; Bednarik, M.; Stanek, M. The Eect of Irradiation on Mechanical and Thermal Properties of Selected Types of Polymers. *Polymers* **2018**, *10*, 158. [CrossRef]
19. Xu, H.; Fang, H.; Bai, J.; Zhang, Y.; Wang, Z. Preparation and Characterization of High-Melt-Strength Polylactide with Long-Chain Branched Structure through γ-Radiation-Induced Chemical Reactions. *Ind. Eng. Chem. Res.* **2014**, *53*, 1150–1159. [CrossRef]
20. Kodal, M.; Wis, A.A.; Ozkoc, G. The mechanical, thermal and morphological properties of γ-irradiated PLA/TAIC and PLA/OvPOSS. *Radiat. Phys. Chem.* **2018**, *153*, 214–225. [CrossRef]

21. Shasteen, C.; Choy, Y.B. Controlling Degradation Rate of Poly(lactic acid) for Its Biomedical Applications. Biomed. *Eng. Lett.* **2011**, *1*, 163–167.
22. Heffner, K.H. Radiation Induced Degradation Pathways for Poly (Methyl Methacrylate) and Polystyrene Polymers as Models for Polymer Behavior in Space Environments. Ph.D. Thesis, University of South Florida, Tampa, FL, USA, 17 November 2003.
23. Paajanen, A.; Sipilä, K. *Modelling Tools for the Combined Effects of Thermal and Radiation Ageing in Polymeric Materials*; Research Report, VTT-R-00102-17; Technical Research Centre of Finland: Espoo, Finland, 2015; p. 15.
24. Ghosal, S.; Mukhopadhyay, M.; Ray, R.; Tarafdar, S. Competitive scission and cross linking in a solid polymer electrolyte exposed to gamma irradiation: Simulation by a fractal model. *Phys. A Stat. Mech. its Appl.* **2014**, *400*, 139–150. [CrossRef]
25. Ghosal, S.; Ray, R.; Mukhopadhyay, M.; Tarafdar, S. Fractal model of polymer electrolyte to investigate influence of high energy irradiation in molecular distribution and molar mass average. *Ionics* **2017**, *23*, 2747–2759. [CrossRef]
26. Saha, M.; Mukhopadhyay, M.; Ray, R.; Ballabh, T.K.; Tarafdar, S. Influence of gamma irradiation on the molecular weight distribution and related properties of a polymer: Study through simulation and experiment. *Model. Simul. Mater. Sci. Eng.* **2014**, *23*, 25003. [CrossRef]
27. Sun, S.H.; Xie, X.; Wu, F. Analysis of nuclear radiation damage effect on typical materials. *IOP Conf. Series: Mater. Sci. Eng.* **2020**, *770*, 012050. [CrossRef]
28. Dole, M. *The Radiation Chemistry of Macromolecules*, 2nd ed.; Academic Press: Cambridge, MA, USA, 1973.
29. Madou, M.J. *Fundamentals of Microfabrication: The Science of Miniaturization*, 2nd ed.; CRC Press: Boca Raton, FL, USA, 2002.
30. Souza, F.C.; Souza, R.F.; Moraes, Â.M. Incorporation and release kinetics of alpha-bisabolol from PCL and Chitosan/guar gum membranes. *Braz. J. Chem. Eng.* **2016**, *33*, 453–467. [CrossRef]
31. Rohman, G.; Pettit, J.J.; Isaure, F.; Cameron, N.R.; Southgate, J. Influence of the physical properties of two-dimensional polyester substrates on the growth of normal human urothelial and urinary smooth muscle cells in vitro. *Biomaterials* **2007**, *28*, 2264–2274. [CrossRef]
32. Garkhal, K.; Verma, S.; Jonnalagadda, S.; Kumar, N. Fast degradable poly(L-lactide-co-ε-caprolactone) microspheres for tis-sue engineering: Synthesis, characterization, and degradation behavior. *J. Polym. Sci. Part A Polym. Chem.* **2007**, *45*, 2755–2764. [CrossRef]
33. Şen, M.; Uzun, C.; Kantoğlu, Ö.; Erdoğan, S.M.; Deniz, V.; Güven, O. Effect of gamma irradiation conditions on the radiation-induced degradation of isobutylene–isoprene rubber. *Nucl. Instrum. Methods B* **2003**, *208*, 480–484. [CrossRef]
34. Quynh, T.M.; Mai, H.H.; Lan, P.N. Stereocomplexation of low molecular weight poly(L-lactic acid) and high molecular weight poly(D-lactic acid), radiation crosslinking PLLA/PDLA stereocomplexes and their characterization. *Radiat. Phys. Chem.* **2013**, *83*, 105–110. [CrossRef]
35. Vazquez-Vazquez, F.C.; Chanes-Cuevas, O.A.; Masuoka, D.; Alatorre, J.A.; Chavarria-Bolaños, D.; Vega-Baudrit, J.R.; Serrano-Bello, J.; Alvarez-Perez, M.A. Biocompatibility of Developing 3D-Printed Tubular Scaffold Coated with Nanofibers for Bone Applications. *J. Nanomater.* **2019**, *2019*, 1–13. [CrossRef]
36. Jose, M.V.; Thomas, V.; Dean, D.R.; Nyairo, E. Fabrication and characterization of aligned nanofibrous PLGA/Collagen blends as bone tissue scaffolds. *Polymer* **2009**, *50*, 3778–3785. [CrossRef]
37. Kodama, Y.; Machado, L.D.; Giovedi, C.; de Lima, N.B.; Nakayama, K. Investigation of irradiated biodegradable blends by FTIR and wide-angle X-ray diffraction. *Nukleonika* **2009**, *54*, 107–113.
38. Lee, J.S.; Chae, G.S.; Khang, G.; Kim, M.S.; Cho, S.H.; Lee, H.B. The effect of gamma irradiation on PLGA and release behavior of BCNU from PLGA wafer. *Macromol. Res.* **2003**, *11*, 352–356. [CrossRef]
39. Di Foggia, M.; Corda, U.; Plescia, E.; Taddei, P.; Torreggiani, A. Effects of sterilisation by high-energy radiation on biomedical poly-(ε-caprolactone)/hydroxyapatite composites. *J. Polym. Sci. Part A Mater. Med.* **2010**, *21*, 1789–1797. [CrossRef]
40. Pietrzak, W.S. Effects of ethylene oxide sterilization on 82:18 PLLA/PGA copolymer craniofacial fixation plates. *J. Craniofac. Surg.* **2010**, *21*, 177–181. [CrossRef]
41. Wang, L.; Xin, B. Poly(L-lactide-co-ε-caprolactone) Nanofiber Morphology Control and Influence of Properties. *J. Phys. Conf. Ser.* **2021**, *1790*, 012062. [CrossRef]
42. Nugroho, P.; Mitomo, H.; Yoshii, F.; Kume, T. Degradation of poly(l-lactic acid) by γ-irradiation. *Polym. Degrad. Stab.* **2001**, *72*, 337–343. [CrossRef]
43. Chen, C.; Xiong, Y.; Jiang, W.; Wang, Y.; Wang, Z.; Chen, Y. Experimental and Numerical Simulation of Biodegradable Stents with Different Strut Geometries. *Cardiovasc. Eng. Technol.* **2019**, *11*, 36–46. [CrossRef]
44. Macha, I.J.; Ben-Nissan, B.; Vilchevskaya, E.N.; Morozova, A.S.; Abali, B.E.; Müller, W.H.; Rickert, W. Drug Delivery From Polymer-Based Nanopharmaceuticals—An Experimental Study Complemented by Simulations of Selected Diffusion Processes. Front. *Bioeng. Biotechnol.* **2019**, *7*, 37.
45. Kashi, S.; Gupta, R.K.; Bhattacharya, S.N.; Varley, R.J. Experimental and simulation study of effect of thickness on performance of (butylene adipate-co-terephthalate) and poly lactide nanocomposites incorporated with graphene as stand-alone electromagnetic interference shielding and metal-backed microwave absorbers. *Compos. Sci. Technol.* **2020**, *195*, 108186.
46. Gluszewski, W. The use of gas chromatography for the determination of radiolytic molecular hydrogen, the detachment of which initiates secondary phenomena in the radiation modification of polymers. *Polimery* **2019**, *64*, 697–702. [CrossRef]

MDPI
St. Alban-Anlage 66
4052 Basel
Switzerland
Tel. +41 61 683 77 34
Fax +41 61 302 89 18
www.mdpi.com

Materials Editorial Office
E-mail: materials@mdpi.com
www.mdpi.com/journal/materials

www.ingramcontent.com/pod-product-compliance
Lightning Source LLC
Chambersburg PA
CBHW062035220926
51850CB00010B/2561
9783036567525